CW00970380

Nestlé Nutrition Institute Workshop Series

Vol. 84

Next-Generation Nutritional Biomarkers to Guide Better Health Care

Editors

Emmanuel E. Baetge Lausanne, Switzerland

Anil Dhawan London, UK

Andrew M. Prentice London, UK

Nestlé Nutrition Institute

Nestec Ltd., 55 Avenue Nestlé, CH–1800 Vevey (Switzerland)
S. Karger AG, P.O. Box, CH–4009 Basel (Switzerland) www.karger.com

Library of Congress Cataloging-in-Publication Data

Names: Nestlé Nutrition Workshop (84th : 2014 : Lausanne, Switzerland), author. | Baetge, Emmanuel E., editor. | Dhawan, Anil, editor. | Prentice, Andrew, editor.
Title: Next-generation nutritional biomarkers to guide better health care / editors, Emmanuel E. Baetge, Anil Dhawan, Andrew M. Prentice.
Description: Basel ; New York : Karger, [2016] | Series: Nestlé Nutrition Institute workshop series, ISSN 1664-2147 ; vol. 84 | Includes bibliographical references and indexes.
Identifiers: LCCN 2015037463| ISBN 9783318055986 (hard cover : alk. paper) | ISBN 9783318055993 (electronic version)
Subjects: | MESH: Biological Markers--Congresses. | Nutritional Physiological Phenomena--Congresses. | Forecasting--methods--Congresses. | Health Status Indicators--Congresses.
Classification: LCC R857.M3 | NLM QW 541 | DDC 610.284--dc23 LC record available at http://lccn.loc.gov/2015037463

Printed on acid-free and non-aging paper
ISBN 978–3–318–05598–6
e-ISBN 978–3–318–05599–3
ISSN 1664–2147
e-ISSN 1664–2155

Basel · Freiburg · Paris · London · New York · Chennai · New Delhi · Bangkok · Beijing · Shanghai · Tokyo · Kuala Lumpur · Singapore · Sydney

Contents

VII **Preface**

X **Foreword**

XIII **Contributors**

Methodologies: Global Epidemiology

1 **Systems-Level Nutrition Approaches to Define Phenotypes Resulting from Complex Gene-Environment Interactions**
Kaput, J. (Switzerland)

15 **Applications of Nutritional Biomarkers in Global Health Settings**
Prentice, A.M. (UK)

25 **Next-Generation Biomarkers of Health**
van Ommen, B.; Wopereis, S. (The Netherlands)

35 **Bioinformatics: Novel Insights from Genomic Information**
Hancock, R.E.W. (Canada)

47 **Summary on Methodologies: Global Epidemiology**
Prentice, A.M. (UK)

Applications/End Users

49 **Biomarkers in Pediatric Liver Disease**
Kyrana, E.; Fitzpatrick, E.; Dhawan, A. (UK)

59 **Next-Generation Biomarkers for Iron Status**
Drakesmith, H. (UK)

71 **The Search for Biomarkers of Long-Term Outcome after Preterm Birth**
Parkinson, J.R.C.; Hyde, M.J.; Modi, N. (UK)

81 **Beyond Cholesterol – New Cardiovascular Biomarkers**
Mangge, H. (Austria)

89 **Summary on Applications/End Users**
Dhawan, A. (UK)

Future Horizons

91 **Stratified Medicine: Maximizing Clinical Benefit by Biomarker-Driven Health Care**
Singh, S. (USA)

103 **The Gut Microbiome, Its Metabolome, and Their Relationship to Health and Disease**
Wu, G.D. (USA)

111 **The Scientific Challenge of Expanding the Frontiers of Nutrition**
Rezzi, S.; Solari, S.; Bouche, N.; Baetge, E.E. (Switzerland)

121 **Summary on Future Horizons**
Baetge, E.E. (Switzerland)

123 **Subject Index**

For more information on related publications, please consult the NNI website: www.nestlenutrition-institute.org

Preface

There are diverse definitions of what constitutes a biomarker for health and currently used formulae driving health care decisions frequently combine a broad range. For instance prediction of a patient's 10-year risk of a coronary event might combine measures of age, body mass index, waist circumference, blood pressure, and blood lipids. Such predictions have sufficient predictive value to have been adopted to set qualifying criteria for long-term antihypertensive and statin treatments. The purpose of the Nestlé Nutrition Institute's 84th Workshop on 'Next-Generation Nutritional Biomarkers to Guide Better Health Care', was to take stock of recent developments unleashed by the 'omics' sciences and make some over-the-horizon predictions of how these exciting new technologies could be harnessed to drive advances in nutrition-related health care.

The potential power of new technologies is breathtaking but, like a wild stallion, the challenges of taming these powers and directing them to useful ends is daunting. Even the most fervent advocates of the new technologies admit that, to date, they have promised more than they have delivered; so it is timely to take stock and ask whether we have reached a take-off point.

Jim Kaput sets the scene with a characteristically revolutionary vision of how 21st century systems thinking must supplant 20th century reductionism and will require not only the new technologies of high-throughput metabolomics, proteomics, genomics, and epigenomics, but also a whole new approach to experimental design in order to tease out how each individual responds to nutritional change. This theme is further developed by *Ben van Ommen* and *Suzan Wopereis*, who base their arguments around a concept of 'systems flexibility' essential to the maintenance of optimal health. Foods and nutrients can play positive or negative roles in determining systems flexibility and their effects can be assessed by interrogating the complex patterns of, for instance, stress response biomarkers. Methodologies for extracting novel biological insights from the terabytes of information that can be generated by omics methods are described by *Robert E.W. Hancock*, whose laboratory has been at the forefront of developing supervised and unsupervised methods for interpreting innate immune and inflammatory responses to infection. The bioinformatic principles involved are representative of the challenges that are faced with next-generation nutritional biomarkers.

Subsequent papers pull back a little from these prophetic visions of what the future holds and concentrate on some of the less complex, but nonetheless state-of-the-art biomarker developments in particular settings or disease states. *Andrew M. Prentice* takes a pragmatic view of the numerous challenges still faced in providing laboratory methods, preferably point-of-care ones, for assessing nutrient deficiencies in poor populations from low-income countries. One example where recent discoveries show great promise for delivering a much needed point-of-care diagnostic is covered by *Hal Drakesmith*. He describes the range of classical, but individually problematic, biomarkers for iron status and concludes that hepcidin, the master regulator of iron metabolism, might yield an advanced biomarker because its levels represent an innate molecular integration of iron status in several body stores and balance the need for iron against the threat of infections that might be promoted by iron administration.

A similar integrative view of biomarker development is presented by *Anil Dhawan* et al. who describe recent advances in the identification of biomarkers of pediatric liver disease, including those corresponding to inflammation, cell death, fibrosis, and the development of malignancy.

In many instances the challenge is to assess existing nutrient status or the state of a diet-related disease so that remedial action can be taken at the individual, group, or even country level. A separate challenge, and the holy grail of many biomarker endeavors, is to discover biomarkers that are predictive of the likelihood of future disease, which could be prevented if detected at an early stage. To this end, *Neena Modi* et al. review the available methods for the prognostic assessment of the likelihood of a preterm birth as well as biomarkers of possible downstream sequelae. They urge the establishment of large prospective cohorts linked with omics methods to advance what is presently a very limited portfolio of methods.

Harald Mangge leads us beyond cholesterol with a discussion of novel cardiovascular biomarkers. He concentrates on the need for early detection of vulnerable atherosclerotic plaques that are undetectable by conventional imaging and classical risk factors. He concludes that, although novel markers exist they may require longitudinal studies of variations within patients in order to detect the onset of a likely critical event. The concept of developing computerized longitudinal monitoring of complex biometric signatures within individual patients is likely to be a future trend and will require prodigious data storage systems within future health services.

The series concludes with three papers taking us into that future. *Sharat Singh* shares his commercial experience of developing tools for the diagnosis and clinical staging of inflammatory bowel diseases to illustrate how such tools can assist in developing stratified and personalized approaches to patient care. *Gary D. Wu* tackles the huge challenge of how to integrate knowledge on the composition and activity of our gut microbiome into clinically tractable information that will benefit future strategies for the prevention and treatment of numerous disorders. Finally, *Emmanuel E. Baetge* et al. develop a theme that has run through the series: namely that the best health care focuses not on the post hoc treatment of disorders once they have occurred but on optimizing human health throughout the life course. Their vision is to optimize the complex nutrition and health 'interactome' by harnessing new nutrient assessment and diagnostic capabilities to drive algorithms that would in turn drive nutritional innovations in the food system from farm to plate.

Although the challenges of reaching some of these space age visions remain formidable this volume confirms that they are now within reach thanks to advances in omic methods and mathematical and programming techniques for extracting actionable information. The future is bright.

Emmanuel E. Baetge
Anil Dhawan
Andrew M. Prentice

Foreword

Biomarkers are the backbones of our daily medical decision making. They indicate the malfunction of organ systems and efficacy of therapeutic interventions. However, only in a few areas related to human nutrition and metabolism, biomarkers play important roles to predict health and functional outcome, and are routinely used in clinical practice. Parameters indicating the lipid status/metabolism which predict long-term cardiovascular risk and efficacy of targeted interventions have been included in routine patient management for a long time. Biomarkers of lipid status have also shown the limitations of our present nutritional intervention strategies. Iron deficiency and overload can be precisely diagnosed and managed by biomarkers, and it has recently been shown that the best 'biomarker' to guide treatment may be the physiological 'determinant' of iron utilization (hepcidin). Interestingly, out of the 'big four' global nutritional deficiencies (vitamin A, zinc, iodine, and iron) which are associated with significant morbidity and mortality in humans, three still cannot be precisely diagnosed by employing (biochemical) biomarkers. Therefore, intervention strategies are mostly still targeted at population level and biomarkers play a limited role in research and decision making. Satisfactory pediatric nutritional biomarkers of outcome must be predictive of later functional health and ideally remain stable over the period from infancy to childhood and adult life. Current traditional biomarkers such as anthropometry and blood pressure are indices that best fulfill those criteria. They are important to monitor long-term health of children who were born with low birth weight in terms of malnutrition or obesity.

New biomarkers which have recently been developed by employing high-throughput metabolomic, proteomic, and genomic technologies indicate that individuals are genetically and biochemically distinct. Our epigenome and metabolome can be influenced by dietary, lifestyle, and environmental factors, which contribute to the heterogeneity observed in humans. Therefore, the risk factors determined for populations cannot be applied to the individual, one has to accept 'individual variability'. New biomarkers which indicate the individual risk or benefit must not neglect the complexity of foods, lifestyle, and metabolic processes that contribute to health or disease and are significant challenges for personalizing dietary advice for healthy or diseased individuals.

The 84th Nestlé Nutrition Institute (NNI) Workshop focused on values and limitations of traditional nutritional biomarkers and opportunities of new biomarkers. NNI would like to thank the three Chairmen, Prof. Emmanuel E. Baetge (Switzerland), Prof. Anil Dhawan (UK), and Prof. Andrew Prentice, (UK) for their challenging and interesting program together and all speakers for their significant contributions. We would also thank the Nestlé Health Science Institute and the Nestlé Research Centre for cooperation and support.

Prof. Ferdinand Haschke
Board Member
Nestlé Nutrition Institute
Vevey, Switzerland

Dr. Natalia Wagemans
Head
Nestlé Nutrition Institute
Vevey, Switzerland

84th Nestlé Nutrition Institute Workshop
Lausanne, September 23–25, 2014

Contributors

Chairpersons & Speakers

Prof. Emmanuel E. Baetge
Director, Nestlé Institute of Health Sciences
Quartier de l'Innovation
CH–1015 Lausanne
Switzerland
E-Mail Edward.Baetge@rd.nestle.com

Prof. Anil Dhawan
Director, Pediatric Liver, GI and Nutrition Centre
Clinical Director, Child Health
King's College Hospital
London SE5 9RS
UK
E-Mail Anil.Dhawan@nhs.net

Prof. Hal (Alexander) Drakesmith
Radcliffe Department of Medicine
John Radcliffe Hospital
University of Oxford
Level 6, West Wing
Headington, Oxford OX3 9DU
UK
E-Mail hdrakes@hammer.imm.ox.ac.uk

Prof. Robert E.W. Hancock
Director, Centre for Microbial Diseases and Immunity Research
University of British Columbia
Room 232
2259 Lower Mall Research Station
Vancouver, BC V6T 1Z4
Canada
E-Mail bob@hancocklab.com

Dr. Jim Kaput
Senior Expert
Nutrition and Metabolic Health Unit
Nestle Institute of Health Sciences
Bldg H
EPFL Campus
CH–1015 Lausanne
Switzerland
E-Mail James.Kaput@rd.nestle.com

Prof. Harald Mangge
Clinical Institute for Medical and Chemical Laboratory Diagnosis
Medical University of Graz
Auenbruggerplatz 15
AT–8036 Graz
Austria
E-Mail harald.mangge@medunigraz.at

Prof. Neena Modi
Section of Neonatal Medicine
Department of Medicine
Imperial College London,
Chelsea and Westminster Hospital Campus
369 Fulham Road
London SW10 9NH
UK
E-Mail n.modi@imperial.ac.uk

Prof. Andrew M. Prentice
MRC International Nutrition Group
Nutrition and Public Health Intervention Research Unit
London School of Hygiene and Tropical Medicine
Keppel Street
London, WC1E 7HT
UK
E-Mail Andrew.Prentice@lshtm.ac.uk

Prof. Sharat Singh
Vice President, Research and Chief Scientific Officer
Prometheus Laboratories Inc.
9410 Carroll Park Drive
San Diego, CA 92121
USA
E-Mail ssingh@prometheuslabs.com

Dr. Ben van Ommen
Principal Scientist/Program Director
Systems Biology TNO
TNO Innovation for Life
Utrechtseweg 48
NL–3704 HE Zeist
The Netherlands
E-Mail ben.vanommen@tno.nl

Prof. Gary D. Wu
Ferdinand G. Weisbrod Professor in Gastroenterology
Division of Gastroenterology
Department of Medicine
University of Pennsylvania
915 Biomedical Research Building
421 Curie Boulevard
Philadelphia, PA 19104
USA
E-Mail gdwu@Mail.med.upenn.edu

Participants

Ying Gao/China
Xu Nicy/China
Lichen Yang/China
Xiaoguang Yang/China
Xiaodan Yu/China
Wendy Hamilton/Dominican Republic
Jorge Palacios-Rosales/Guatemala
Alfredo Larossa Haro/Mexico
Rafael Ponce de León Barajas/Mexico
Enrique Romero Velarde/Mexico
Salavador Villalpando Carrion/Mexico
Michael Affolter/Switzerland
Laurent Ameye/Switzerland
Nicolas Bouche/Switzerland
Laura Camurri/Switzerland
Ralf Crabbé/Switzerland
Nanda De Groot/Switzerland
Patrick Descombes/Switzerland
Julie Deuquet/Switzerland
Stéphane Duboux/Switzerland
Jane Durga/Switzerland
Christophe Fuerer/Switzerland
Jean-Philippe Godin/Switzerland
Michaela Hoehne/Switzerland
Valérie Marquardt/Switzerland
Yery Antonio Mendoza/Switzerland
Angus Moodycliffe/Switzerland
Frédéric Raymond/Switzerland
Serge Rezzi/Switzerland
Claudia Roessle/Switzerland
Pierre-Philippe Sagnier/Switzerland
Simona Stan/Switzerland
Robert Carlson/USA
Sheri Volger/USA

Baetge EE, Dhawan A, Prentice AM (eds): Next-Generation Nutritional Biomarkers to Guide Better Health Care. Nestlé Nutr Inst Workshop Ser, vol 84, pp 1–13, (DOI: 10.1159/000436947)
Nestec Ltd., Vevey/S. Karger AG., Basel, © 2016

Systems-Level Nutrition Approaches to Define Phenotypes Resulting from Complex Gene-Environment Interactions

Jim Kaput

Systems Nutrition and Health, Nestlé Institute of Health Sciences, Lausanne, Switzerland

Abstract
High-throughput metabolomic, proteomic, and genomic technologies have delivered 21st-century data showing that humans cannot be randomized into groups: individuals are genetically and biochemically distinct. Gene-environment interactions caused by unique dietary and lifestyle factors contribute to the heterogeneity in physiologies observed in human studies. The risk factors determined for populations (i.e. the population-attributable risk) cannot be applied to the individual. Developing individual risk/benefit factors in light of the genetic diversity of human populations, the complexity of foods, culture and lifestyle, and the variety in metabolic processes that lead to health or disease are significant challenges for personalizing dietary advice for healthy or diseased individuals.

Introduction

Two of the great advances in biomedical research over the past 100 years were the standardization of experimental designs, specifically the case-control design [1], and the one gene-one polypeptide concept that emanated from the groundbreaking work of Beadle and Tatum [2]. These contributions, in addition to a few other seminal discoveries (e.g. elucidation of the DNA structure), laid the conceptual framework for 20th-century biological research. Much of the research of the last 80 years is what Kuhn [3] called normal science. This type of research activity does not produce new concepts, but rather finds facts to match and better explain the existing framework. Normal science standardizes experimental designs that ultimately are considered mandatory for interpreting data and publishing results. Scientific dogma holds, for example, that the only trustworthy results of human studies are from prospective randomized controlled trials (RCTs) [4, 5]. Similarly, many studies in humans or laboratory animals are reductionistic in analyzing how, for example, a single gene or a single nutrient correlates with some physiological effect [2].

Revolutions can occur when normal science produces experimental data that cannot be explained by existing paradigms [3]. The outcomes of this paradigm shift have profound implications for biomarker identification and development that are necessary to assess nutrition and lifestyle choices for maintaining or improving personal and public health.

Analyzed Heterogeneity Supplants Randomization

In addition to the elaborations of human experimental designs, legal developments occurred over a long period that began with the passage of the US Food, Drug, and Cosmetic Act in 1938. The Cosmetic Act required that new drugs undergo premarket safety evaluation, although the legal mandate to prove efficacy was not enacted until 1962. The Drug Amendment Act specifically required adequate and well-controlled clinical investigations with positive findings from at least two clinical studies [6]. The scientific precedents for these legal actions were built on the success of analyzing the efficacy of antibiotics [6]. Infectious bacteria (e.g. *Vibrio cholera)* are extrinsic agents that affect virtually all humans. Hence, the average response between the treated group and the control group in RCTs can provide proof of the efficacy of the treatment. In contrast, drugs, nutrition, and lifestyle choices become intrinsic factors in that they interact with and affect internal physiological functions.

Unlike extrinsic agents, intrinsic factors may be metabolized by or interact with physiological systems. Humans (and all species) not only show biochemical individuality [7] at homeostasis, but may also respond differently to drug or food chemicals [8–10]. Differential responses occur because each individual is unique [11] and genetic variations may be differently expressed in response to nutrients and other factors. Heterogeneity within a species is, of course, the fundamental basis by which natural selection acts to produce evolutionary changes. The statistical result of RCTs is the population-attributable risk (PAR), defined as the number (or proportion) of cases that would not occur *in a population* if the factor were eliminated [12] – they are not individual risk factors [13]. While PARs are applicable for large effect sizes and extrinsic agents, the utility of PARs is diminished by the heterogeneity of many individual genetic makeups added to the calculations, the latent effects of many interacting environmental factors (dietary chemicals and activity levels), and the resulting individuality of metabolic responses [14–16]. The human system, which is comprised of a person's genetic makeup, microbiome, diet, lifestyle, and resulting physiology, acts on (metabolizes) treatments, and the treatments differentially alter the physiology depending upon the individual. The result of this heterogeneity can be (and often is) that the distribution of metabolic measures (or responses) in the case group can overlap with the distribution of metabolic measures (or responses) in the control group [17].

Since the pregenomic era had limited methodological tools to analyze human variation at the genetic or physiological level, randomization was essential for 20th-century biomedical research. However, genome sequencing has not only demonstrated genetic individuality [reviewed in ref. 11, 18, 19] but can now be used to completely characterize genetic makeups of study participants. Recent data [20, 21] suggest that each person differs from others and the reference genomes by about 3.5 million single nucleotide polymorphisms (SNPs), almost 1,000 large copy number variants (CNVs), and large numbers of insertions and deletions. Different levels of DNA methylation and therefore epigenetic regulation have also been demonstrated [22–24]. Variation in the (epi)genetic makeup may express itself in variation in RNA abundance [25, 26] and therefore levels of proteins and enzymes. To add to this complexity, physiological variability is influenced by the human microbiome [27, 28], the combination of all microorganisms that reside on the skin, in saliva, in the oral mucosa, in the conjunctiva, in the gastrointestinal tract, and in the vagina. Each of these factors alone or in combination could alter the level of a biomarker.

Nutrition and lifestyle not only alter the expression of information encoded in the genome [29], they can also modify the epigenome [30] and the microbial composition [31]. No experiments are needed to demonstrate that individuals have heterogeneous dietary intakes and physical activity levels. The result of interactions among (epi)genetic, microbiome, nutritional, and environmental factors is that humans have heterogeneous metabolomic profiles [32–34] in health and disease states. Physiological variability has been recognized for centuries and summarized in the modern pregenomic era by Williams [7] in 1956 in a book entitled *Biochemical Individuality*, and by nutritionists of the 1960s and 1970s [35, 36]. Many (but not all) of the previously unmeasurable, molecular characteristics of individuals can now be analyzed.

Medical practitioners and health professionals treat individuals and not population groups [35]. Individual risk factors are unknown, even though the concepts of personal determinants of health were taught by Hippocrates and Galen centuries ago. In addition to Williams' [7] treatise in 1956, others also expressed the need and approach to analyze individuals rather than groups [35–40]. More recently, we [41–44] and others [45–47] have been promoting or using n-of-1 aggregation and analysis methods [45, 48]. The concept of n-of-1 studies is that each person serves as his or her own control. Physiological assessments are usually done before and after a treatment [39] or intervention. The results of each trial (that is, from one individual) can then be aggregated for statistical analysis [45]. For example, we aggregated results from data obtained at homeostasis to analyze group average differences between males and females [48], but also found that each individual varies in micronutrient levels. This variation could then be used to associate patterns of plasma protein levels and variations in the genetic makeup [41].

Most -omes will remain incompletely characterized because they are so complex in composition and because they vary in time and space. Nevertheless, the incomplete molecular data sets can be analyzed to generate a model that predicts new markers to test in remaining samples of the experiment, or as new markers in the follow-on experiment in an iterative process of refining the model. Regardless of the new methodological and analytical approaches, the quantitative postgenomic data demonstrating human metabolomic and physiological heterogeneity should have profound impacts on the design of human research studies and, specifically, the validity of RCTs for determining optimal nutrition or drug treatments. Hence, even though not all physiological variables can be analyzed, randomization becomes unnecessary since 21st-century technologies produce enough data to more completely characterize each individual.

Systems Thinking Supplants Reductionism

Biochemical research over the past 100 years identified pathway maps consisting of individual components and sequential enzymatic reactions [2] for metabolizing carbohydrates, lipids, and amino acids, for synthesizing and degrading nucleic acids, for producing energy, for transporting metabolites, proteins, and nucleic acids, and for regulating hormone status. Studies of intracellular signaling in cellular growth and death pathways extended metabolic maps to signal transduction and gene-regulatory functions. However, even these regulatory networks were reduced to 2D maps of interconnecting components. Many of these pathway maps were derived from studies of disease rather than health processes, which is not reflected in those representations. Nevertheless, the beauty and simplicity of these maps leads to the illusion of explanatory depth [49] – a concept that elegant visual figures can appear as reality rather than that they represent simple one-to-one relationships that are frozen in time and space.

Pathway maps were and continue to be highly influential in guiding the design of experiments to elucidate the mechanisms of physiological disorders such as cancer, diabetes, obesity, cardiovascular disease, and other chronic diseases. However, relying on pathway maps to design experiments explicitly reduces the complexity of a phenotype to single (or a few) reactions in a pathway regardless of the experimental model (i.e. transgenic animals, cell model systems, or humans). For example, high fasting glucose levels, a hallmark of type-2 diabetes, may be caused by overproduction of glucose through gluconeogenesis in the liver, increased absorption of dietary carbohydrates in the intestine, decreased production of insulin from the pancreas, or insulin resistance [44, 50, 51]. Understanding one of these pathways does not provide comprehensive understanding of type-2 diabetes, which results from a complex set of interactions between multiple genes and multiple environmental factors.

Many researchers in the nutrition community also adopted reductionistic experimental designs by analyzing the effects of a single nutrient on complex phenotypes including health and disease states. Since most nutrients are ingested in small amounts, exposure to safe and small increases in the test nutrient over baseline intake requires months for a physiological or anthropometric effect large enough to be measured. A nutrient, of course, cannot be given alone but must be consumed in the background of a normal diet, which is not only complex, but can vary between individuals. Nota bene – individuals deficient in a single nutrient can be successfully treated by a reductionistic approach: worldwide incidence rates of rickets (vitamin D), beriberi (thiamine), scurvy (vitamin C), pellagra (niacin), and night blindness (vitamin A) have all been reduced by single-nutrient interventions in malnourished individuals [52–55]. However, the impact of subclinical undernourishment and the needs of populations in different environments and with diverse cultural, genetic, and agricultural histories are unknown.

The focus on individual genetic variants or the identification of independent environmental factors [55–57] needs to evolve toward a more comprehensive analysis of nutrient intakes, environmental and lifestyle factors, genetic makeups, and physiology [58]. High-throughput omic technologies have generated paradigm-shifting data that challenge the conceptual basis not only of RCTs but also experimental reductionism.

Systems-level designs and analyses of high-dimensional data have been reported for studies of cardiovascular disease, obesity, diabetes, nutrition, drug intake, toxicology, immunology, gut microbiota, medicine, health care, and health disparities [59–67]. These systems-level reports analyzed the patterns within one or at most two data type(s), i.e. the interaction network of metabolites or between metabolites and proteins. In most of these publications, the system was closed since variables such as diet intake, lifestyle, or other environmental variables were not measured or included in the analysis. Excluding external factors that influence internal biological processes generates at best an incomplete system, and likely an inaccurate understanding of the interactions between environment and genetic makeup. From a practical standpoint, such a design misses an opportunity to identify modifiable factors that influence health. Biological processes occur in open systems [68], and ex vivo factors including nutrients and other naturally occurring chemicals in food can alter biochemical processes and signaling networks occurring within the organism [29]. Several recent reports included dietary intake variables as a part of omic-based systems [69–71] or genomic analysis [41, 48, 72]. These studies are consistent with a paradigm shift from a focus on individual genetic variants or the identification of single independent environmental factors [56–58] toward a more comprehensive analysis of nutrient intakes, environmental and lifestyle factors, genetic makeups, and physiology [58].

Systems, Heterogeneity, and Biomarkers

The reality of human heterogeneity and the view of physiology as a system are not abstractions, but have consequences for the design of experiments [42] and interpretation of data [41, 48], including the discovery, verification, and validation of biomarkers (table 1). Metabolites (e.g. cholesterol, calcium, and homocysteine), proteins (e.g. insulin), supramolecular complexes (LDL or HDL particles), or conditions (e.g. blood pressure) have long been used as clinical biomarkers. The omics revolution accelerated biomarker research of disease [73–78] and health [79]. Genomic technologies based on SNPs in candidate genes [80], CNVs [81], transcriptomics [71, 82, 83], and microRNAs [84, 85] provide a new means to explore multiple biomarkers for pathologies. However, the majority of biomarker discovery studies were conducted using RCTs and hence result in PAR rather than individual risk. Most individual biomarkers and additive combinations of SNPs identified by genome-wide association studies explain between 1% (obesity) and 20% (ankylosing spondylitis) of the phenotype of interest [86]. Geneticists refer to the remaining variability as missing heritability [87, 88], apparently ignoring that many phenotypes are the result of gene-gene (epistatic [89]), epigenetic (e.g. DNA methylation [90]), and gene-environment interactions [91].

Systems thinking is starting to influence nutritional [41, 48, 69–72, 91, 92] and immunological [66] research, although significant challenges remain in adapting and applying these advances to human studies [93]. More specifically, metabolic processes underlying health, disease initiation, disease progression, and disease

Table 1. Definitions used for biomarkers

Term	Definition	References
Accuracy	Closeness of agreements between the value of the measured and the true concentration in that sample	98
Biomarker	A characteristic that is objectively measured and evaluated as an indicator of normal biologic processes, pathogenetic processes, or pharmacologic responses to a therapeutic intervention	96, 100, 101
Clinical end point	A characteristic or variable that reflects how a patient feels or functions, or how long a patient survives	96, 100, 101
Diagnostic predictability	Ability of the test to predict the presence or absence of a disease for a given test result and determined by calculating the positive and negative predictive values Positive predictive values are the proportion of patients with positive test results who have the disease, while negative predictive values are the proportion of patients with negative test results who do not have the disease	98
Differentiation	Differentiation of efficacy or safety of a drug within the same class	77
Homeostasis	The steady states of systems and physiologies in an organism – the constancy of the internal environment in two separate states: sleep and awake	102, 103
Hormesis	A dose-response relationship phenomenon characterized by low-dose stimulation and high-dose inhibition	104
Nutritional phenotype	Defined and integrated set of genetic, proteomic, metabolomic, functional, and behavioral factors that form the basis for the assessment of nutritional status	105
Prognostic factor	Individuals with disease have biomarkers that are predictive over time and require evidence for validity Comparative and equivalent to risk factors	106
Reliability/ repeatability	Ability to replicate tests to yield the same results under the same measurement conditions	98
Reproducibility	Describes measurements performed under different conditions	98
Risk factor	Individuals without disease have biomarkers that are predictive over time and require evidence for validity Comparative and equivalent to prognostic factors	106
Sensitivity	Proportion of individuals who test positive for a given biomarker: reflects the true positive rate	98, 99
Specificity	Proportion of individuals without symptoms who yield negative results: reflects false-positive rate	98, 99
Stratification	Select patients to increase likelihood of therapeutic success	77
Surrogate end point (or outcome)	A biomarker intended to substitute for a clinical end point A clinical investigator uses epidemiologic, therapeutic, pathophysiologic, or other scientific evidence to select a surrogate end point to predict clinical benefit, harm, or lack of benefit or harm	96, 101, 106
Trueness	Closeness of agreement between the average value of different samples and the true concentration value	98

See also http://www.genomicglossaries.com/content/Biomarkers.asp.

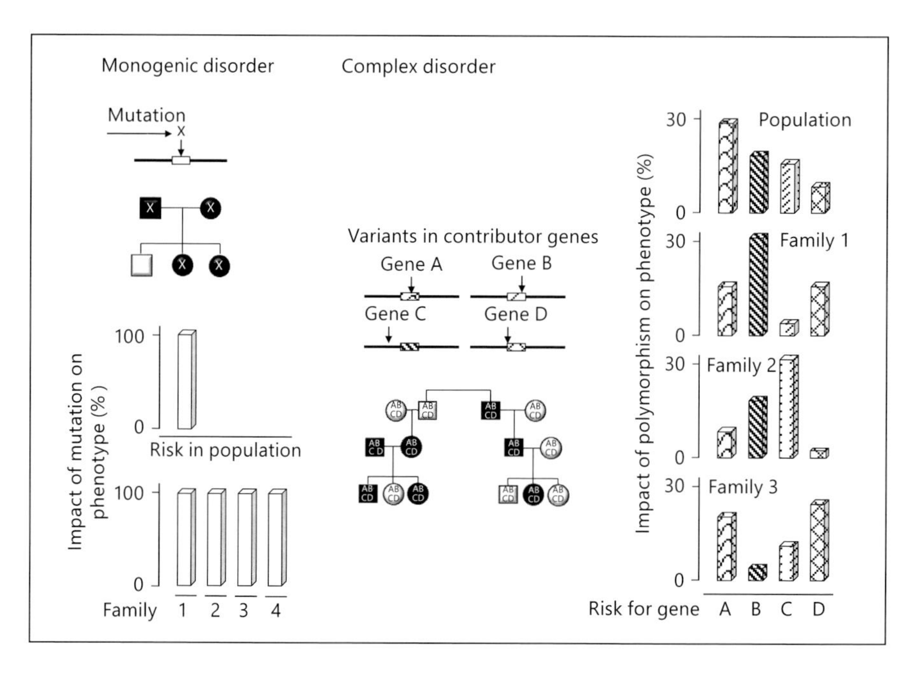

Fig. 1. Monogenic versus complex disorders. Monogenic diseases are caused by (in this case, an autosomal dominant) mutations in a single gene and hence the genetic risk is the same for each family or individual who inherits it. Pedigrees can be used to trace the presence of the mutation. Affected individuals are shown with a black symbol with white X (mutation) in the pedigree. Complex disorders (or phenotypes) are caused by genes (e.g. A, B, C, and D) having one or more variations (SNPs, CNVs, insertions, and deletions) and epigenetic factors interacting with other genes (epistasis) and with environmental factors. Pedigrees reveal no Mendelian inheritance pattern due to the effect size of each variant in the particular genetic background of the individual (represented by the size of the letters in the pedigree). PAR factors (figure labeled population) for each variant in each gene (A, B, C, and D, with the pattern of the box matched to the risk factor in the graphs) are the average of each variation in the tested population. The phenotype in different families or individuals may be impacted differently by each of these variants (family 1, 2, and 3 in the graphs). Some genes that predispose individuals to disease might have minor or no effect in some families/individuals (gene D, family 2). This figure was adapted from Peltonen and McKusik [94].

outcome are linked networks whose robustness and individual variability are poorly understood. A likely reason is that biomarkers are discovered and analyzed individually. For example, a genome-wide association study statistically corrects p values of each SNP individually for multiple comparisons. A systems view of interacting processes is that patterns of markers will better indicate the state of the system than single markers. For most complex phenotypes, a biomarker will have a different effect size in each individual depending on gene-gene, epigenetic, and gene-environment interactions (fig. 1). This concept was explained for genetic contributions to complex disease phenotypes by Peltonen and McKusick [94].

An added complexity for biomarker discovery is the dynamic nature of health and disease processes. Since most of these diseases have a late onset, biomarkers are typically associated with surrogate end points which, ideally, would be equivalent to the clinical end point [95, 96]. However, the same clinical end point can result from imbalances in different organs, pathways, and genes, which has been acknowledged as a confounder in the development of single biomarkers [97, 98]. A case in point is type-2 diabetes mellitus, as explained earlier. Multiple biomarkers will be needed for each chronic disease [98] and likely for stages of each disease. Subsets of the pathways and genes for each surrogate end point may be differentially affected by long-term exposure to diet and other environmental factors [44, 51], which may alter the reliability, reproducibility, trueness, and diagnostic predictability of their measurement.

Health-related biomarkers have an added level of complexity since many of the parameters will have small deviations from normal values in different individuals, and added strength must be gained from the use of combinations of single biomarkers (biomarker profiles based on additivity or multiplicity of effects). So far, these profiles have been (mis)used as statistical observation for group differences without a mechanistic or functional linkage, and usually in the absence of knowledge about long-term nutritional status. The biologically (and nutritionally) relevant profile (or interactome-representing biomarker) will only emerge if these relationships are established through quantitative assessment of the overarching processes [79]. Based on these concepts, each disease or health biomarker in a panel will have a probability of diagnostic predictability and, more importantly, the panel itself will have a probability associated with its predictive efficacy. Developing these biomarker panels will require novel experimental designs for discovery [42], verification, and validation. The era of personalized health care and nutrition will not emerge until genetic heterogeneity, environmental complexity, and physiological variability are taken into account when conducting human biomedical research. However, the result of including nutrition in research will result in evidence-based knowledge that can be applied to improving human health through nutrition and lifestyle choices.

Acknowledgment

The author thanks Martin Kussmann and Laura Camurri for reviewing this paper.

Disclosure Statement

The author works for the for-profit Nestle Institute of Health Sciences.

References

1 Fisher RA: The Design of Experiments. London, Oliver & Boyd, 1935.

2 Beadle G, Tatum E: Genetic control of biochemical reactions in neurospora. Proc Natl Acad Sci U S A 1941;27:499–506.

3 Kuhn TS: The Structure of Scientific Revolutions: 50th Anniversary Edition. Chicago, University of Chicago Press, 1962, p 218.

4 Begg C, Cho M, Eastwood S, et al: Improving the quality of reporting of randomized controlled trials. The CONSORT Statement. JAMA 1996;276:637–639.

5 Schulz KF, Altman DG, Moher D, CONSORT Group: CONSORT 2010 statement: updated guidelines for reporting parallel group randomised trials. BMJ 2010; 340:c332.

6 Junod SW: FDA and clinical drug trials: a short history; in Davies M, Kerimani F (eds): A Quick Guide to Clinical Trials. Washington, Bioplan, 2008, pp 25–55.

7 Williams RP: Biochemical Individuality: The Basis for the Genetotrophic Concept. New Canaan, Keats Publishing, 1956.

8 Pellis L, van Erk MJ, van Ommen B, et al: Plasma metabolomics and proteomics profiling after a postprandial challenge reveal subtle diet effects on human metabolic status. Metabolomics 2012;8:347–359.

9 Crandall JP, Shamoon H, Cohen HW, et al: Post-challenge hyperglycemia in older adults is associated with increased cardiovascular risk profile. J Clin Endocrinol Metab 2009;94: 1595–1601.

10 Krug S, Kastenmüller G, Stückler F, et al: The dynamic range of the human metabolome revealed by challenges. FASEB J 2012;26: 2607–2619.

11 Olson MV: Human genetic individuality. Annu Rev Genomics Hum Genet 2012;13: 1–27.

12 Rockhill B, Newman B, Weinberg C: Use and misuse of population attributable fractions. Am J Public Health 1998;88:15–19.

13 Kaput J: Nutrigenomics research for personalized nutrition and medicine. Curr Opin Biotechnol 2008;19:110–120.

14 Shin S-Y, Fauman EB, Petersen A-K, et al: An atlas of genetic influences on human blood metabolites. Nat Genet 2014;46:543–550.

15 Bouatra S, Aziat F, Mandal R, et al: The human urine metabolome. PLoS One 2013;8:e73076.

16 Robinette SL, Holmes E, Nicholson JK, Dumas ME: Genetic determinants of metabolism in health and disease: from biochemical genetics to genome-wide associations. Genome Med 2012;4:30.

17 Meigs JB, Shrader P, Sullivan LM, et al: Genotype score in addition to common risk factors for prediction of type 2 diabetes. N Engl J Med 2008;359:2208–2219.

18 Mardis ER: A decade's perspective on DNA sequencing technology. Nature 2011;470: 198–203.

19 Ng PC, Kirkness EF: Whole genome sequencing. Methods Mol Biol 2010;628:215–226.

20 Durbin RM, Abecasis GR, Altshuler DL, et al: A map of human genome variation from population-scale sequencing. Nature 2010; 467:1061–1073.

21 Sudmant PH, Kitzman JO, Antonacci F, et al: Diversity of human copy number variation and multicopy genes. Science 2010;330:641–646.

22 Jaffe AE, Irizarry RA: Accounting for cellular heterogeneity is critical in epigenome-wide association studies. Genome Biol 2014;15: R31.

23 Liang L, Cookson WOC: Grasping nettles: cellular heterogeneity and other confounders in epigenome-wide association studies. Hum Mol Genet 2014;23:R83–R88.

24 Campión J, Milagro FI, Martínez JA: Individuality and epigenetics in obesity. Obes Rev 2009;10:383–392.

25 Kliebenstein DJ: Quantification of variation in expression networks. Methods Mol Biol 2009;553:227–245.

26 Montgomery SB, Dermitzakis ET: From expression QTLs to personalized transcriptomics. Nat Rev Genet 2011;12:277–282.

27 Benson AK, Kelly SA, Legge R, et al: Individuality in gut microbiota composition is a complex polygenic trait shaped by multiple environmental and host genetic factors. Proc Natl Acad Sci U S A 2010;107:18933–18938.

28 ElRakaiby M, Dutilh BE, Rizkallah MR, et al: Pharmacomicrobiomics: the impact of human microbiome variations on systems pharmacology and personalized therapeutics. OMICS 2014;18:402–414.

29 Kaput J, Rodriguez RL: Nutritional genomics: the next frontier in the postgenomic era. Physiol Genomics 2004;16:166–177.

30 Burdge GC, Lillycrop KA: Bridging the gap between epigenetics research and nutritional public health interventions. Genome Med 2010;2:80.

31 Flint HJ: The impact of nutrition on the human microbiome. Nutr Rev 2012;70(suppl 1): S10–S13.

32 Zivkovic AM, German JB: Metabolomics for assessment of nutritional status. Curr Opin Clin Nutr Metab Care 2009;12:501–507.

33 Suhre K, Shin SY, Petersen AK, et al: Human metabolic individuality in biomedical and pharmaceutical research. Nature 2011;477: 54–60.

34 Illig T, Gieger C, Zhai G, et al: A genome-wide perspective of genetic variation in human metabolism. Nat Genet 2010;42:137–141.

35 Editorial: Disposition to obesity. Br Med J 1964;ii:1543–1544.

36 Young VR, Scrimshaw NS: Genetic and biological variability in human nutrient requirements. Am J Clin Nutr 1979;32:486–500.

37 Guyatt G, Sackett D, Adachi J, et al: A clinician's guide for conducting randomized trials in individual patients. CMAJ 1988;139:497–503.

38 Guyatt G, Jaeschke R: N-of-1 randomized trials – where do we stand? West J Med 1990; 152:67–68.

39 Guyatt GH, Keller JL, Jaeschke R, et al: The n-of-1 randomized controlled trial: clinical usefulness. Our three-year experience. Ann Intern Med 1990;112:293–299.

40 Editorial: Embracing patient heterogeneity. Nat Med 2014;20:689.

41 Morine M, Monteiro J, Wise C, et al: Genetic associations with micronutrient levels identified in immune and gastrointestinal networks. Genes Nutr 2014;9:408.

42 Kaput J, Morine M: Discovery-based nutritional systems biology: developing N-of-1 nutrigenomic research. Int J Vitam Nutr Res 2012;82:333–341.

43 Kussmann M, Kaput J: Translational genomics. Appl Transl Genomics 2014, pp 1–5, http://www.academia.edu/7689433/ATG_2014_KUSSMANN_KAPUT_Translational_Genomics.

44 Kussmann M, Morine MJ, Hager J, et al: Perspective: a systems approach to diabetes research. Front Genet 2013;4:205.

45 Nikles J, Mitchell GK, Schluter P, et al: Aggregating single patient (n-of-1) trials in populations where recruitment and retention was difficult: the case of palliative care. J Clin Epidemiol 2011;64:471–480.

46 Bacchetti P, Deeks SG, McCune JM: Breaking free of sample size dogma to perform innovative translational research. Sci Transl Med 2011;3:87ps24.

47 Kelley JM, Kaptchuk TJ: Group analysis versus individual response: the inferential limits of randomized controlled trials. Contemp Clin Trials 2010;31:423–428.

48 Monteiro JP, Wise C, Morine MJ, et al: Methylation potential associated with diet, genotype, protein, and metabolite levels in the Delta Obesity Vitamin Study. Genes Nutr 2014;9:403.

49 Rozenblit L, Keil F: The misunderstood limits of folk science: an illusion of explanatory depth. Cogn Sci 2002;26:521–562.

50 Stumvoll M: Control of glycaemia: from molecules to men. Minkowski Lecture 2003. Diabetologia 2004;47:770–781.

51 Kaput J, Noble J, Hatipoglu B, et al: Application of nutrigenomic concepts to type 2 diabetes mellitus. Nutr Metab Cardiovasc Dis 2007;17:89–103.

52 Allen L, de Benoist B, Dary O, Hurrell R: Guidelines on food fortification with micronutrients, 2006, http://www.who.int/nutrition/publications/micronutrients/9241594012/en/.

53 Boy E, Mannar V, Pandav C, et al: Achievements, challenges, and promising new approaches in vitamin and mineral deficiency control. Nutr Rev 2009;67(suppl 1):S24–S30.

54 Save the Children: Nutrition in the First 1,000 Days. State of the World's Mothers 2012. Westport, Save the Children, 2012.

55 Patel CJ, Bhattacharya J, Butte AJ: An Environment-Wide Association Study (EWAS) on type 2 diabetes mellitus. PLoS One 2010;5: e10746.

56 Patel CJ, Chen R, Butte AJ: Data-driven integration of epidemiological and toxicological data to select candidate interacting genes and environmental factors in association with disease. Bioinformatics 2012;28:i121–i126.

57 Patel CJ, Cullen MR, Ioannidis JPA, Butte AJ: Systematic evaluation of environmental factors: persistent pollutants and nutrients correlated with serum lipid levels. Int J Epidemiol 2012;41:828–843.

58 Kaput J, van Ommen B, Kremer B, et al: Consensus statement – understanding health and malnutrition through a systems approach: the ENOUGH program for early life. Genes Nutr 2014;9:378.
59 Kalupahana NS, Moustaid-Moussa N: Overview of symposium 'Systems Genetics in Nutrition and Obesity Research'. J Nutr 2011; 141:512–514.
60 Meng Q, Mäkinen V-P, Luk H, Yang X: Systems biology approaches and applications in obesity, diabetes, and cardiovascular diseases. Curr Cardiovasc Risk Rep 2013;7:73–83.
61 Slikker W Jr, Paule MG, Wright LK, et al: Systems biology approaches for toxicology. J Appl Toxicol 2007;27:201–217.
62 Auffray C, Chen Z, Hood L: Systems medicine: the future of medical genomics and healthcare. Genome Med 2009;1:2.
63 Kleemann R, Bureeva S, Perlina A, et al: A systems biology strategy for predicting similarities and differences of drug effects: evidence for drug-specific modulation of inflammation in atherosclerosis. BMC Syst Biol 2011;5:125.
64 Roux AVD: Complex systems thinking and current impasses in health disparities research. Am J Public Health 2011;101:1627–1634.
65 Gardy JL, Lynn DJ, Brinkman FSL, Hancock REW: Enabling a systems biology approach to immunology: focus on innate immunity. Trends Immunol 2009;30:249–262.
66 Afacan NJ, Fjell CD, Hancock REW: A systems biology approach to nutritional immunology – focus on innate immunity. Mol Aspects Med 2012;33:14–25.
67 Karlsson FH, Nookaew I, Petranovic D, Nielsen J: Prospects for systems biology and modeling of the gut microbiome. Trends Biotechnol 2011;29:251–258.
68 von Bertalanffy L: The theory of open systems in physics and biology. Science 1950; 111:23–29.
69 Morine MJ, McMonagle J, Toomey S, et al: Bi-directional gene set enrichment and canonical correlation analysis identify key diet-sensitive pathways and biomarkers of metabolic syndrome. BMC Bioinformatics 2010; 11:499.
70 Morine MJ, Tierney AC, van Ommen B, et al: Transcriptomic coordination in the human metabolic network reveals links between n-3 fat intake, adipose tissue gene expression and metabolic health. PLoS Comput Biol 2011;7: e1002223.
71 Morine MJ, Toomey S, McGillicuddy FC, et al: Network analysis of adipose tissue gene expression highlights altered metabolic and regulatory transcriptomic activity in high-fat-diet-fed IL-1RI knockout mice. J Nutr Biochem 2013;24:788–795.
72 Nettleton JA, McKeown NM, Kanoni S, et al: Interactions of dietary whole-grain intake with fasting glucose- and insulin-related genetic loci in individuals of European descent: a meta-analysis of 14 cohort studies. Diabetes Care 2010;33:2684–2691.
73 Frantzi M, Bhat A, Latosinska A: Clinical proteomic biomarkers: relevant issues on study design & technical considerations in biomarker development. Clin Transl Med 2014;3:7.
74 Suhre K: Metabolic profiling in diabetes. J Endocrinol 2014;221:R75–R85.
75 Beger R: A review of applications of metabolomics in cancer. Metabolites 2013;3:552–574.
76 Drucker E, Krapfenbauer K: Pitfalls and limitations in translation from biomarker discovery to clinical utility in predictive and personalised medicine. EPMA J 2013;4:7.
77 Dayon L, Núñez Galindo A, Corthesy J, et al: Comprehensive and scalable highly automated MS-based proteomic workflow for clinical biomarker discovery in human plasma. J Proteome Res 2014, Epub ahead of print.
78 Marshall J, Bowden P, Schmit JC, Betsou F: Creation of a federated database of blood proteins: a powerful new tool for finding and characterizing biomarkers in serum. Clin Proteomics 2014;11:3.
79 van Ommen B, Keijer J, Heil SG, Kaput J: Challenging homeostasis to define biomarkers for nutrition related health. Mol Nutr Food Res 2009;53:795–804.
80 Kraft P, Wacholder S, Cornelis MC, et al: Beyond odds ratios – communicating disease risk based on genetic profiles. Nat Rev Genet 2009;10:264–269.
81 Fanciulli M, Petretto E, Aitman TJ: Gene copy number variation and common human disease. Clin Genet 2010;77:201–213.

82 Sagaya FM, Hurrell RF, Vergères G: Postprandial blood cell transcriptomics in response to the ingestion of dairy products by healthy individuals. J Nutr Biochem 2012;23:1701–1715.
83 Afman LA, Müller M: Human nutrigenomics of gene regulation by dietary fatty acids. Prog Lipid Res 2012;51:63–70.
84 Leidinger P, Backes C, Deutscher S, et al: A blood based 12-miRNA signature of Alzheimer disease patients. Genome Biol 2013;14:R78.
85 Wang K, Zhang S, Marzolf B, et al: Circulating microRNAs, potential biomarkers for drug-induced liver injury. Proc Natl Acad Sci U S A 2009;106:4402–4407.
86 Visscher PM, Brown MA, McCarthy MI, Yang J: Five years of GWAS discovery. Am J Hum Genet 2012;90:7–24.
87 Eichler EE, Flint J, Gibson G, et al: Missing heritability and strategies for finding the underlying causes of complex disease. Nat Rev Genet 2010;11:446–450.
88 Lee SH, Wray NR, Goddard ME, Visscher PM: Estimating missing heritability for disease from genome-wide association studies. Am J Hum Genet 2011;88:294–305.
89 Steffens M, Becker T, Sander T, et al: Feasible and successful: genome-wide interaction analysis involving all 1.9 × 10(11) pair-wise interaction tests. Hum Hered 2010;69:268–284.
90 Edwards SL, Beesley J, French JD, Dunning AM: Beyond GWASs: illuminating the dark road from association to function. Am J Hum Genet 2013;93:779–797.
91 Zheng J-S, Arnett DK, Lee Y-C, et al: Genome-wide contribution of genotype by environment interaction to variation of diabetes-related traits. PLoS One 2013;8:e77442.
92 Merched AJ, Chan L: Nutrigenetics and nutrigenomics of atherosclerosis. Curr Atheroscler Rep 2013;15:328.
93 Norheim F, Gjelstad IMF, Hjorth M, et al: Molecular nutrition research: the modern way of performing nutritional science. Nutrients 2012;4:1898–1944.
94 Peltonen L, McKusick VA: Genomics and medicine. Dissecting human disease in the postgenomic era. Science 2001;291:1224–1229.
95 De Gruttola VG, Clax P, DeMets DL, et al: Considerations in the evaluation of surrogate endpoints in clinical trials: summary of a National Institutes of Health workshop. Control Clin Trials 2001;22:485–502.
96 Strimbu K, Tavel JA: What are biomarkers? Curr Opin HIV AIDS 2010;5:463–466.
97 Rifai N, Gillette MA, Carr SA: Protein biomarker discovery and validation: the long and uncertain path to clinical utility. Nat Biotechnol 2006;24:971–983.
98 Maruvada P, Srivastava S: Joint National Cancer Institute-Food and Drug Administration workshop on research strategies, study designs, and statistical approaches to biomarker validation for cancer diagnosis and detection. Can Epidemiol Biomarkers Prev 2006;15:1078–1082.
99 Biomarkers Definitions Working Group: Biomarkers and surrogate endpoints: preferred definitions and conceptual framework. Clin Pharmacol Ther 2001;69:89–95.
100 Weir CJ, Walley RJ: Statistical evaluation of biomarkers as surrogate endpoints: a literature review. Stat Med 2006;25:183–203.
101 Cannon WB: Organization for physiological homeostasis. Physiol Rev 1929;9:399–431.
102 Recordati G, Bellini TG: A definition of internal constancy and homeostasis in the context of non-equilibrium thermodynamics. Exp Physiol 2004;89:27–38.
103 Calabrese EJ, Baldwin LA: Hormesis: the dose-response revolution. Annu Rev Pharmacol Toxicol 2003;43:175–197.
104 Zeisel SH, Freake HC, Bauman DE, et al: The nutritional phenotype in the age of metabolomics. J Nutr 2005;135:1613–1616.
105 Lassere MN, Johnson KR, Boers B, et al: Definitions and validation criteria for biomarkers and surrogate endpoints: development and testing of a quantitative hierarchical levels of evidence schema. J Rheumatol 2007;34:607–615.
106 Bonassi S, Neri M, Puntoni R: Validation of biomarkers as early predictors of disease. Mutat Res 2001;480–481:349–358.

Baetge EE, Dhawan A, Prentice AM (eds): Next-Generation Nutritional Biomarkers to Guide Better Health Care. Nestlé Nutr Inst Workshop Ser, vol 84, pp 15–23, (DOI: 10.1159/000436948)
Nestec Ltd., Vevey/S. Karger AG., Basel, © 2016

Applications of Nutritional Biomarkers in Global Health Settings

Andrew M. Prentice

MRC International Nutrition Group, London School of Hygiene and Tropical Medicine, London, UK

Abstract

In global health settings, there are three generic areas that require reliable biomarkers of nutritional status and function. Population surveillance needs to identify key nutrient deficiencies (or excesses) to monitor progress towards elimination of nutritional imbalances and to stratify populations into groups especially 'at risk' to whom public health resources can be focused. Clinical interventions need biomarkers to help identify disease pathways, to assist in targeting nutrient prescriptions, and to avoid potential harm (e.g. in the case of iron). Discovery science requires biomarkers in many domains, but especially in the study of nutrient-gene interactions and regarding the effects of nutritional status on the epigenome. Each of these applications imposes different constraints on the methodology though in all cases the optimum biomarker would have high sensitivity and specificity, would capture variation of functional significance, and would be cheap and easy to apply. These attributes are hard to achieve, and recent progress towards next-generation biomarkers, though holding much promise, has not yet delivered significant breakthroughs in the global health setting. Recent efforts to overcome these problems by two initiatives (BOND and INSPIRE) are highlighted as exemplars of a route map to progress.

Introduction

Following significant economic progress in many regions of the world, and consequent improvements in access to better quality diets, the residual global public health burden of nutrient (especially micronutrient) deficiencies is now concentrated on the world's poorest populations that are largely confined to sub-Saharan Africa and South Asia. In such regions, there are multiple overlapping applications for biomarkers of nutritional status ranging from individual clinical diagnosis, through group screening for targeted interventions, to population surveillance.

Screening for gross nutritional sufficiency of protein-energy supply tends to still focus on simple anthropometric indicators assessing stunting (height-for-age z score), underweight (weight-for-age z score), wasting (weight-for-height z score), and mid-upper arm circumference. Assessment of protein status still relies on crude indicators based on plasma protein concentrations (e.g. albumin) which are insensitive to all except gross protein deficiency. In these aspects, there has been little notable movement towards next-generation indices of nutritional status. One exception is the use of noninvasive measures of body composition, most importantly for assessing body fat, lean tissue, and muscle mass. In advanced research facilities, dual X-ray absorptiometry and air displacement plethysmography (e.g. BODPOD® and PEAPOD®) are being directed towards the further understanding of the etiology and later sequelae of early growth failure, but for most low-income clinics or field applications, bioimpedance assessment is the only really practical method.

Assessment of micronutrient status also remains a major challenge with few significant breakthroughs that have been adequately validated in recent years. The challenges are manifold. First, micronutrients can be divided into type-1 and type-2 micronutrients [1, 2]. For type-1 micronutrients, the category that includes all vitamins and most minerals, physical growth continues in the face of deficiency, and hence tissue levels are depleted, but characteristic clinical signs only become visible at extreme levels of deficiency. For type-2 micronutrients (e.g. zinc and protein), growth slows rapidly and hence tissue levels tend to be maintained making detection of deficiency very challenging. Second, for many nutrients, the levels in blood (the customary biopsy tissue) are homeostatically maintained by reserves in the liver or other tissues. Thus, measuring circulating levels of vitamin A, for instance, provides only a crude measure of vitamin A status and is more useful at the population level than at the individual level. Third, the circulating levels of many micronutrients are profoundly altered by inflammation raising challenges as to how to correct for these effects especially in low-income settings where infections are common. Such problems over the validity of existing biomarkers for nutrient status create a problem for developing new tests; namely, that there is no gold standard against which to reference new techniques.

Ideally, next-generation biomarkers would be based on functional tests. Some tests already exist. For instance the erythrocyte glutathione reductase activation coefficient test assesses the percent saturation of erythrocyte glutathione reductase with its riboflavin-derived cofactor flavin adenine dinucleotide [3]. This has been shown to robustly correlate with the activities of other flavo-enzymes in other tissues thus providing a comprehensive status assessment [3]. There are a few other examples of such functional indicators, but others are still required.

It appears that the discovery of the iron-regulatory hormone hepcidin provides a step forward in assessing iron status, and consequently Drakesmith [this vol., pp. 59–69] allocated a full paper to this issue in this workshop. We believe that the opposing transcriptional regulators of hepcidin synthesis (iron deficiency and inflammation) reflect evolutionary pressures to optimize iron status in the face of possible infectious threats, and hence hepcidin can form the basis of a next-generation point-of-care diagnostic that could indicate '*ready and safe to receive iron*' [4, 5].

The BOND and INSPIRE Initiatives

The Gates Foundation/National Institutes of Health/National Institute of Child Health and Human Development (NICHD)-sponsored programs BOND (Biomarkers of Nutrition for Development) [6] and INSPIRE (Inflammation and Nutritional Science for Programs/Policies and Interpretation of Research Evidence) [7], led by Dr. Daniel Raiten at NICHD, have pulled together experts to address the challenges highlighted above. Outputs, in both published and interactive web-based formats, from each initiative provide excellent resources describing the range of biomarkers currently available and in development.

The BOND program has started by addressing 6 key nutrients: vitamins A and B_{12}, iron, iodine, folate, and zinc. Table 1 (reproduced from BOND) shows the high-level summary of biomarkers available for vitamin A. Examination of this table reveals many of the challenges and roadblocks faced across the wider portfolio of biomarkers for micronutrient status. For instance, the most widely used measures (retinol-binding protein, serum retinol, and breast milk retinol) provide poor measures of status at the individual level, frequently fail to respond to intervention in predictable ways, and are affected by inflammation. Measures that provide a better index of the status of an individual require mass spectrometry, and hence are costly and have long lag times to the production of results, thus making them only applicable in research settings. Physiological function tests are either hard to implement (e.g. dark adaptation) or insensitive (e.g. self-reported night blindness). These limitations in assessing vitamin A status are replicated to a greater or lesser extent across the other 5 nutrients, but the BOND summaries provide investigators

Table 1. High-level summary of biomarkers for vitamin A status from the BOND initiative[1]

Biomarkers	Type	Use	Utility
Serum retinol-binding protein	Status (deficiency)	Population	Not released from the liver when retinol is limited Used as a proxy for serum retinol to identify vitamin A deficiency
Serum/ plasma retinol	Status	Population	Most commonly used biomarker Correlates with the prevalence and severity of xerophthalmia and may change in response to interventions
Relative dose response	Status	Population Individual	Based on hepatic accumulation of retinol-binding protein during vitamin A depletion Requires blood sample before and after an oral retinyl ester dose
Modified relative dose response	Status	Population Individual	More responsive than serum retinol Qualitatively identifies low or adequate liver vitamin A reserves
Retinol isotope dilution	Status, marker of excess	Population	Most sensitive test to measure vitamin A status and intervention impact on vitamin A reserves Minimally invasive and accurate
Breast milk retinol	Status, exposure	Population	Good indicator of vitamin A status in areas where breastfeeding is common until at least 6 months of age Impacted by many factors
Retinyl esters	Status, marker of excess	Population Individual	Validated qualitative measure of hypervitaminosis A May be confounded by liver disease at the individual level
Dark adaptation	Function	Population (small scale) Individual	Dark-adapted final threshold is inversely and sensitively correlated with serum vitamin A levels in low to deficient ranges
Electro-retinography	Function	Population Individual	Measures the bioelectrical response of the retina to a flash of light Invasive and nonsuitable for children
Pupillary threshold testing	Function	Population Groups of individuals	Inversely correlates with serum vitamin A values in low to deficient ranges and the concentration of vitamin A in the retina Noninvasive, can be used in field conditions
Dietary assessment	Exposure	Population Individual (repeated testing)	Qualitative measure of exposure Provides useful information to support biochemical biomarkers Seasonality of fruits and vegetables must be included

[1] Reproduced with permission from BOND (https://www.nichd.nih.gov/global_nutrition/programs/bond/Pages/index.aspx).

at all levels with a roadmap for navigating towards optimally matching methods with applications.

Attempts to design next-generation methodologies are very challenging for a number of reasons. First, the endeavors to develop more precise, reliable, and cost-effective assays and to multiplex these cannot overcome the basic physiological reasons that limit interpretation. Second is the question of a lack of a gold standard reference method against which to calibrate any new methodologies under development. For instance, innovative studies of the plasma proteome have elegantly demonstrated both expected and novel proteins correlated with micronutrient status in Nepalese children [8] but in this, and similar, endeavors it is difficult to escape from the circular logic inherent in comparing one technique to another in the absence of a single gold standard. Perhaps the only way around this will be to use multiple methods each with individual uncertainties and to triangulate between them. Such a process is complex to perform and would need to be implemented across various sex and age groups, and account for physiological variables such as inflammation; this explains why it has rarely, if ever, been achieved.

Biomarkers Employed in Discovery Science Applications in Global Health

There is still much to be learnt about the fundamental relationships between diet, health, and disease in both first- and third-world settings, and by virtue of the greater variations in nutrient intakes in poor populations there is a strong imperative to study such populations to seek new solutions. Sometimes studies conducted in nutritionally marginalized populations can provide insights that could also be applied across better-nourished populations. We provide one such example here to illustrate both the power and the limitations of complex biomarker studies.

Epigenetic changes induced in the very early phases of the life cycle are believed to provide at least a partial explanation for the known linkage between the nutritional and environmental circumstances of a mother's pregnancy and the lifelong health of her children or her children's children [9]. One class of epigenetic changes, namely DNA methylation, requires an adequate supply of methyl groups that in turn is dependent on a series of interlocking metabolic pathways in which nutrients provide both substrates (choline, betaine, methionine, and folic acid) and enzyme cofactors (vitamins B_2, B_6, and B_{12}; fig. 1). Knowledge of intermediary metabolites (homocysteine, cysteine, dimethyl glycine (DMG), S-adenosylmethionine (SAM), and S-adenosylhomocysteine (SAH)) may also help inform how nutrient balances could affect DNA methylation [10].

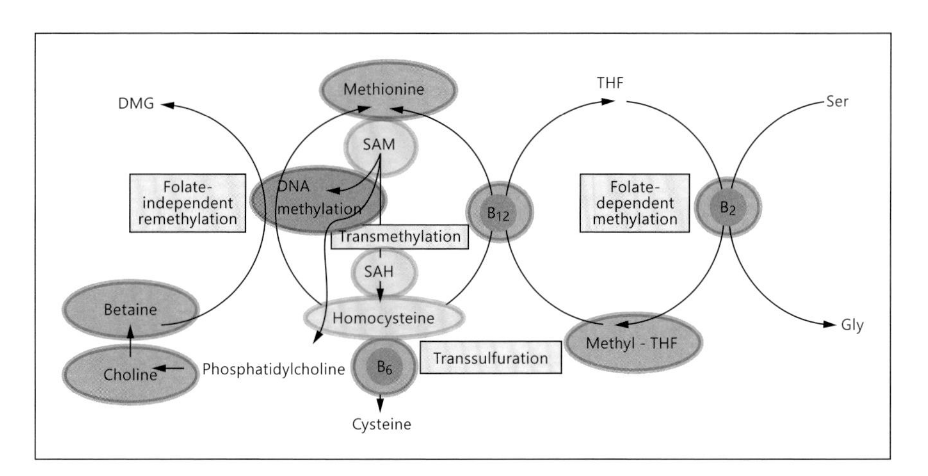

Fig. 1. Biomarkers to assess the methylation pathways and potential in epigenetic studies. The nutrient and intermediary metabolite biomarkers illustrated represent a reduced set of biomarkers (a mini methyl donor metabolome) assessable in blood that are useful in assessing the likely in vivo potential for DNA methylation. THF = Tetrahydrofolate.

We used a set of biomarkers measured during early pregnancy to capture the methyl donor 'minimetabolome' in order to test the hypothesis that deficiencies in these pathways could affect how DNA is remethylated in the very early embryo [11]. We had previously shown that the methylation of metastable epialleles (regions of the genome where methylation patterns are known to be established very soon after conception) is influenced in rural Gambia by whether babies are conceived in the hungry or the harvest season [12]. Surprisingly, the babies born in the harvest season, i.e. when foods are most abundant, showed lower levels of methylation suggesting that the effects were mediated by more complex aspects of nutrient supply than simply the abundance of the diet. We therefore prospectively studied a new cohort of women conceiving in the two seasons [13]. We also studied a group of nonpregnant women with monthly blood samples in order to gain a complete picture of how their diet and methyl donor biomarkers varied throughout the year. Figure 2a shows the annual variations in these nonpregnant women for some of the key intermediary metabolites and shows the large swing in the SAM/SAH ratio, a key determinant of the

Fig. 2. Seasonal differences in biomarkers of the methylation potential in rural Gambian mothers. **a** Seasonal variation in some key methyl donor pathway biomarkers assessed monthly in 30 nonpregnant, nonlactating Gambian women. **b** Heat maps of methyl donor biomarkers assessed in early gestation in 165 women. Each vertical bar represents a subject whose biomarker concentration is expressed as a z score relative to the overall mean. $ActB_{12}$ = Active B_{12}. * $p < 0.05$, ** $p < 0.01$, *** $p < 0.001$. *(For figure see next page.)*

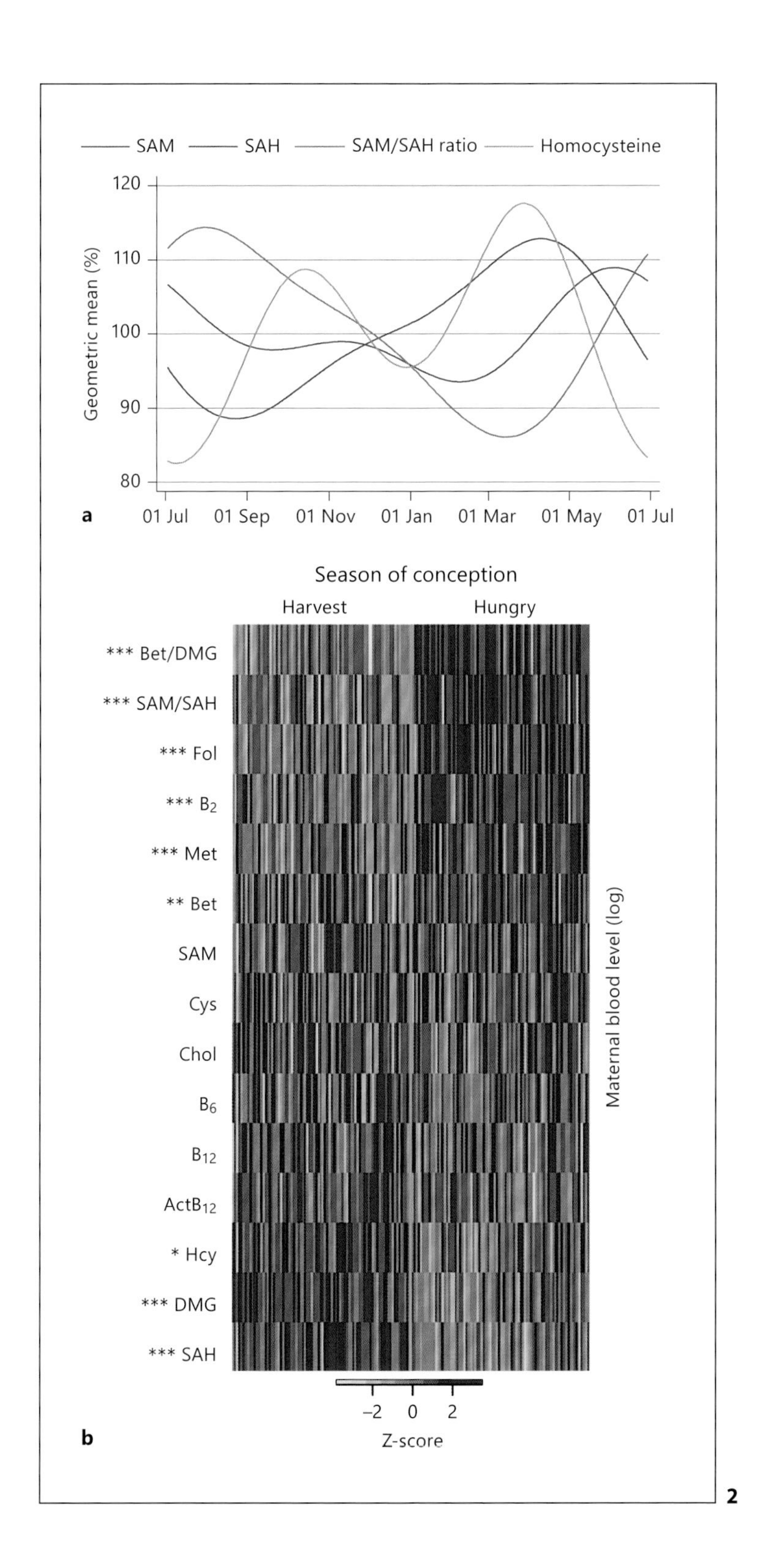
SAM
SAH
SAM/SAH ratio
Homocysteine
Geometric mean (%)
120
110
100
90
80
a
01 Jul
01 Sep
01 Nov
01 Jan
01 Mar
01 May
01 Jul
Season of conception
Harvest
Hungry
*** Bet/DMG
*** SAM/SAH
*** Fol
*** B2
*** Met
** Bet
SAM
Cys
Chol
B6
B12
ActB12
* Hcy
*** DMG
*** SAH
Maternal blood level (log)
−2
0
2
Z-score
b

2

methylation potential. The heat map in figure 2b presents data from 165 women equally divided into those who conceived in the hungry and harvest seasons. It shows that many of the substrates, cofactors, and intermediary metabolites involved in methylation pathways differed significantly between the two seasons, and, as observed previously, there were also significant differences in the methylation of metastable epialleles in the offspring which were significantly correlated with their mothers' biomarker levels. These results provided first-in-human proof that a mother's diet at conception could epigenetically alter her baby's DNA in ways that might have lifelong effects on health.

These are important results but they reveal some significant challenges with respect to the use of biomarkers. First, the assays were costly and time consuming being performed by traditional wet laboratory methodologies. Performing them on a next-generation metabolomics platform would have been preferable if a platform existed that could yet compete with wet laboratory precisions and accuracy, but this goal has yet to be achieved. Second, although the levels of several of the metabolites significantly predicted DNA methylation and all of these were in the correct direction based upon first principles of the biochemical pathways involved, it has remained hard to decipher which nutrient interventions would be most beneficial due to the complexities of the modeling required [14]. Additional analytes and a much larger dataset will be required in the hope of directing the next steps towards developing a cocktail of nutrients to optimize methylation and infant outcomes.

In summary, there remain very significant challenges in designing next-generation biomarker-based methods for global health diagnostics and research. These challenges will be surmountable if sufficient efforts and resources are allocated to finding solutions.

Acknowledgments

A.M.P. is supported by grant No. MCA760-5QX00 to the International Nutrition Group by the UK Medical Research Council (MRC) and the UK Department for International Development (DFID) under the MRC/DFID concordat agreement.

Disclosure Statement

The author has no financial relationships or conflicts of interest relevant to this article to disclose.

References

1 Golden MH: Specific deficiencies versus growth failure: type I and type II nutrients. SCN News 1995;12:10–14.
2 Prentice AM, Gershwin ME, Schaible UE, et al: New challenges in studying nutrition-disease interactions in the developing world. J Clin Invest 2008;118:1322–1329.
3 Bates CJ: Diagnosis and detection of vitamin deficiencies. Br Med Bull 1999;55:643–657.
4 Pasricha SR, Atkinson SH, Armitage AE, et al: Expression of the iron hormone hepcidin distinguishes different types of anemia in African children. Sci Transl Med 2014;6: 235re3.
5 Atkinson SH, Armitage AE, Khandwala S, et al: Combinatorial effects of malaria season, iron deficiency, and inflammation determine plasma hepcidin concentration in African children. Blood 2014;123:3221–3229.
6 Raiten DJ, Namaste S, Brabin B, et al: Executive summary – biomarkers of nutrition for development: building a consensus. Am J Clin Nutr 2011;94:633S–650S.
7 Raiten DJ, Sakr Ashour FA, Ross AC, et al: Inflammation and Nutritional Science for Programs/Policies and Interpretation of Research Evidence (INSPIRE). J Nutr 2015;145: 1S–70S.
8 Cole RN, Ruczinski I, Schulze K, et al: The plasma proteome identifies expected and novel proteins correlated with micronutrient status in undernourished Nepalese children. J Nutr 2013;143:1540–1548.
9 Aiken CE, Ozanne SE: Transgenerational developmental programming. Hum Reprod Update 2014;20:63–75.
10 Dominguez-Salas P, Cox SE, Prentice AM, et al: Maternal nutritional status, C(1) metabolism and offspring DNA methylation: a review of current evidence in human subjects. Proc Nutr Soc 2012;71:154–165.
11 Dominguez-Salas P, Moore SE, Cole D, et al: DNA methylation potential: dietary intake and blood concentrations of one-carbon metabolites and cofactors in rural African women. Am J Clin Nutr 2013;97:1217–1227.
12 Waterland RA, Kellermayer R, Laritsky E, et al: Season of conception in rural Gambia affects DNA methylation at putative human metastable epialleles. PLoS Genet 2010;6: e1001252.
13 Dominguez-Salas P, Moore SE, Baker MS, et al: Maternal nutrition at conception modulates DNA methylation of human metastable epialleles. Nat Commun 2014;5:3746.
14 Scotti M, Stella L, Shearer EJ, Stover PJ: Modeling cellular compartmentation in one-carbon metabolism. Wiley Interdiscip Rev Syst Biol Med 2013;5:343–365.

Baetge EE, Dhawan A, Prentice AM (eds): Next-Generation Nutritional Biomarkers to Guide Better Health Care. Nestlé Nutr Inst Workshop Ser, vol 84, pp 25–33, (DOI: 10.1159/000436949)

Next-Generation Biomarkers of Health

Ben van Ommen · Suzan Wopereis

Department of Systems Biology, The Netherlands Organization for Applied Scientific Research (TNO), Zeist, The Netherlands

Abstract

Current biomarkers used in health care and in nutrition and health research are based on quantifying disease onset and its progress. Yet, both health care and nutrition should focus on maintaining optimal health, where the related biology is essentially differing from biomedical science. Health is characterized by the ability to continuously adapt in varying circumstances where multiple mechanisms of systems flexibility are involved. A new generation of biomarkers is needed that quantifies all aspects of systems flexibility, opening the door to real lifestyle-related health optimization, self-empowerment, and related products and services.

Introduction

Health care does not really focus on maintaining optimal health but rather on curing diseases. A large repertoire of tools, technologies, and treatments has been developed for this purpose, making disease care an enterprise that may soon become too costly. Also, within health care, citizens become patients in the literal sense of the word: patiently undergoing treatments instead of playing an active part in their own health. This needs to change and in theory should be simple as a huge health profit can be achieved if each person would adapt to an optimal lifestyle, including a proper diet, during their lifespan. Reviews suggest that major reductions in obesity, type-2 diabetes (T2D), cardiovascular disease, and cancers could be achieved [1]. Theory and practice differ and we face a multifactorial challenge, spanning economic, social, psychological, and biological aspects. Yet, from a biological viewpoint, a major breakthrough would be achieved if knowledge and technologies would become available that allow to understand and quantify the processes that maintain health. So far, efforts in biomarker development have mostly focused on quantification of disease states or development. This has been relatively easy, as disease biology significantly differs from health biology, and has also been rewarding because the health care economy provided major incentives for such biomarkers (diagnostics). Where diet and nutrition should aim at maintaining optimal health, research in this area was hampered because abundant use was made of 'disease biomarkers'. This created tension in regulation and food industry, and drawbacks for (funding in) nutrition and health research.

What is needed is a complete refocusing of health research, starting with (re) defining health and its related mechanisms, understanding that integrated personalized health optimization strategies are needed, redesigning the methods to quantify health, and from there onwards building a new generation of health biomarkers. These biomarkers have to serve two crucial goals: to report on health improvement or health maintenance instead on disease progression and to empower the individual to achieve this. Each of these aspects is further detailed below.

Systems Flexibility as a Characteristic of Optimal Health

Human health is based on a complex network of interactions between pathways, mechanisms, processes, and organs. Many of these processes have to function in a continuously changing environment (diet, infections, stress, temperature, and exercise, for example) and thus strive to maintain internal homeostasis by adapting to these changes. We call this phenotypic flexibility [2] and realize that disease onset occurs when and where these adaptive processes fail. Importantly, diet plays both a positive and negative role here. Many nutrients serve specifically to optimize these 'flexibility processes' (fig. 1).

Shifting the focus from disease biomarkers towards the development of next-generation biomarkers of (optimal) health needs a different approach to quantify health and different strategies of testing. Health is maintained by a complex interaction of processes, each maintaining 'homeostasis', elasticity, and robustness. This well-orchestrated physiologic machinery (fig. 2) to adapt to the continuously changing environment is termed 'phenotypic flexibility' [2]. A suboptimal health condition becomes apparent under situations of temporary stress, like physical exercise, infections, or mental stress. Also, dietary habits, e.g. excess intake of sugars or fats, present temporary stress to the body. In various systems

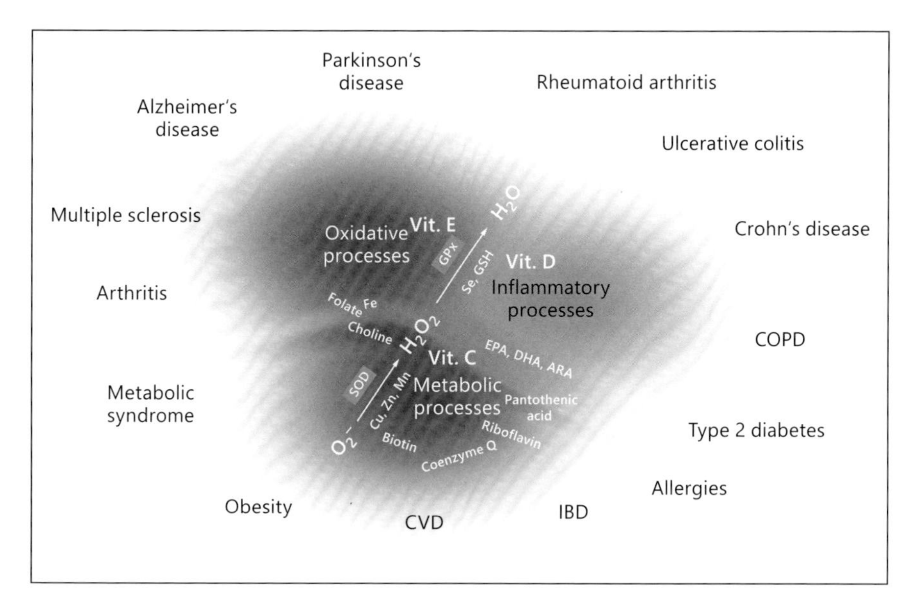

Fig. 1. Many diseases (the outer circle) have a 'lifestyle-related component', yet the mechanisms of disease progression are essentially different from the processes that maintain optimal health. Here, three crucial processes are described, i.e. the capacity to maintain flexibility in metabolic, oxidative, and inflammatory stress. Flexibility in this context is described as the capacity to contain the stress response reactions within 'healthy limits' (both in amplitude and time), and thus the capacity to maintain optimal homeostasis. Interestingly, many essential nutrients function in this area [19]. CVD = Cardiovascular disease; IBD = inflammatory bowel disease; IR = insulin resistance; SOD = superoxide dismutase.

(e.g. transfer of people, goods, finances, and energy), optimal performance is achieved only when logistics and infrastructure function well and are capable of dealing with temporary overload or stress. Disturbances in these systems lead to traffic jams, shortages, or damages. Stress tests are applied to test the flexibility of such systems in unexpected situations. Similarly, proper management of calories and nutrients by our body requires the optimal metabolism and condition of the organs. When this is the case, the body's flexibility is able to cope with temporary distortions, a condition which can be qualified as 'healthy'.

Quantification of Systems Flexibility: Stress Response Biomarkers

Due to a wide variety of reasons (e.g. genetic and epigenetic factors, exposure, diet, stress, and exercise), individuals differ in their 'wiring' of phenotypic flexibility, will react differently to acute and chronic stressors, and develop a personal trajectory of metabolic-inflammatory health and disease. Thus, personalized

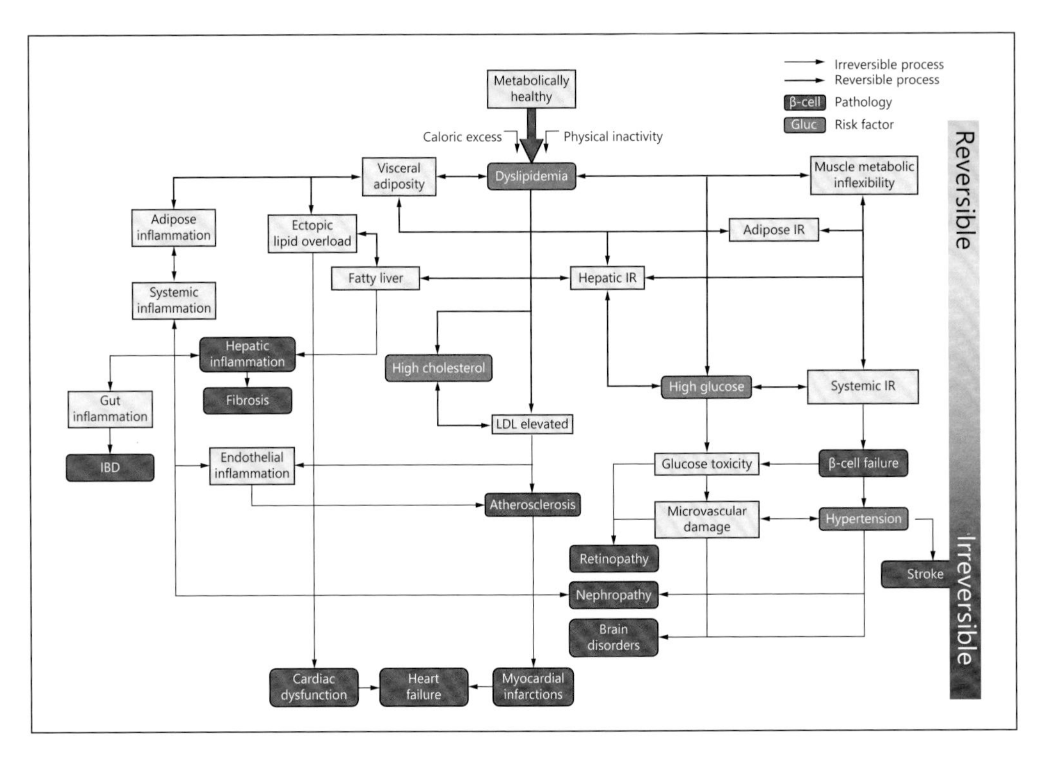

Fig. 2. The system of phenotypic flexibility where multiple processes spanning multiple organs interact in maintaining optimal metabolic-inflammatory health, and where caloric imbalance eventually leads to a series of disorders, developing in a personal manner depending on genotype, previous exposure, and lifestyle (reproduced with authors permission from van Ommen et al. [2]). Gluc = Glucose; IBD = inflammatory bowel disease; IR = insulin resistance; SOD = superoxide dismutase.

diagnosis of the phenotypic flexibility system needs to reveal the 'weak spots' in this flexibility network. For one person, this may be impaired triglyceride storage in adipose tissue resulting in a fatty liver, for another the impaired excretion of VLDL particles from the liver due to a shortage in choline, which results in a fatty liver. A third person may accumulate liver fat due to a shortage of carnitine, which causes inadequate fatty acid oxidation, for example. Each of these processes needs to be diagnosed and require a specific (food-based) therapy [3].

To better quantify these conditions, the development and application of standardized metabolic stress tests have been suggested to quantify health and health effects from diets or treatments [4]. For the different organs and processes shown in figure 2, markers are needed where the 'amplitude' and the 'duration' of disturbance (time needed to get back to homeostatic conditions) are taken as readout. These are 'multibiomarker' panels representing defined and accepted health-related processes that need to be combined with a standardized stress test or challenge test that preferably modulates these defined and accepted health-related processes. In covering the metabolic health arena, flexibility quantification should focus on the 'overarching processes' that are oxidative stress, metabolism, and inflammation (fig. 1), since these processes are important for maintaining health, and disturbances can cause the switch from healthy towards the development of chronic metabolic diseases. Multibiomarker panels will emerge that act as composite descriptors of physiological processes. In the example of vascular health, such a composite biomarker could be composed of flow-mediated dilation, a functional marker of endothelial function and blood pressure, resilience markers for endothelial damage after a metabolic challenge test such as VCAM, ICAM, and E-selectin responses, and total cholesterol or specific single nucleotide polymorphisms related to an increased risk for cardio-metabolic disease development. By combining this information into an integrated readout such as the 'vascular health index' [5] or as a 'health space' [6], a flexibility marker for vascular health can be obtained that has broader value, both for product development and health care. It is important that a standardized stress test or challenge test will be developed that modulates most phenotypic health processes. A standardized challenge test should be characterized in how it modulates the different processes of phenotypic flexibility and how well it differentiates between health states in the sequel from optimal healthy to sub-optimal healthy to diseased, including heterogeneity, subpopulations, and different stages of the disease. Finally, variation in the response to a challenge should be related to established markers of disease or to long-term health outcomes (disease risk/longevity) in cohort studies for their validation.

Systems Interventions – Optimizing Each Process Involved in Systems Flexibility

Systems diseases require systems diagnoses based on quantifying 'phenotypic flexibility' as described in the previous paragraph, revealing the underlying disease cause(s) within a complex network of nonlinear metabolic and inflammatory processes which drive most systems diseases. For optimal phenotypic flexibility, each process needs to function optimally. In T2D, several organs can contribute to disruption of (glucose) metabolism [7]. The degree of insulin sensitivity of the three main organs, pancreas, muscle tissue, and liver, can be assessed by measuring glucose and insulin at 30-min intervals during an oral glucose tolerance test. It is known that the severity of insulin resistance can differ between the various tissues, and that different interventions may have organ-specific effects related to increasing insulin sensitivity [8], as demonstrated by the

Table 1. T2D subgroup (process)-dependent diagnosis/intervention strategies

T2D subgroups (see fig. 2, based on processes involved)	Diagnosis (i.e. parameters of the P4 biopassport)	Potential interventions
(1) Pancreatic β-cell function (impaired IR)	OGTT: I/ΔG and DI(0), PYY, Arg, His, Phe, Val, Leu	β-cell-protective nutrients (MUFA) and drugs (TZD, GLP-1 analogs, and DPP-4 inhibitors)
(2) Muscle IR (decreased glucose uptake)	OGTT: muscle IR index, insulin secretion/IR index, Val, Ile, Leu, γ-glutamyl derivates, Tyr, Phe, Met	PUFA/SFA balance; physical activity; weight loss; TZD (e.g. PPAR-γ)
(3) Hepatic IR with decreased glucose uptake but increased production and release	Hepatic IR index, OGTT, hepatic IS index, ALAT, ASAT, bilirubin, GGT, ALP, CK-18 fragments, lactate, α/β-hydroxybutyrate	Decrease in SFA and n-6 PUFA, and increase in n-3 PUFA; weight loss; metformin; TZD; exenatide (GLP-1 analog); DPP-4 inhibitors
(4) Adipocyte IR and lipotoxicity	Basal adipocyte IR index, FFA platform, glycerol	α-Lipoic acid; PUFA/SFA balance; n-3 FA; chitosan/plant sterols; TZD; acipimox
(5) Gastrointestinal tract (incretin deficiency/resistance)	i.v. GTT vs. OGTT, GLP-1, GIP, glucagon, bile acids	MUFA; dietary fiber (pasta/rye bread); exenatide
(6) Pancreatic α-cell hyperfunction	Fasting plasma glucagon	Glucagon receptor antagonist; exenatide; DPP-4 inhibitors
(7) Chronic low-grade inflammation	CRP, total leukocytes VCAM, ICAM, oxylipids, cytokines	Fish oil/n-3 fatty acids; vitamin C/E; carotenoids; salicylates; TNF-α inhibitors

Currently, 7 processes involved in T2D are identified, and for each of them a biomarker approach to quantify the process, as well as an intervention strategy to optimize/restore health, is suggested. ALP = Alkaline phosphatase; CK = cytokeratin; CRP = C-reactive protein; DDP-4 = dipeptidylpeptidase-4; DI = disposition index; FA = fatty acids; FFA = free FA; GGT = γ-glutamyltransferase; GIP = glucose-dependent insulinotropic polypeptide; GLP = glucagon-like peptide; IR = insulin resistance; IS = insulin sensitivity; MUFA = monounsaturated FA; OGTT = oral glucose tolerance test; PPAR-γ = peroxisome proliferator-activated receptor-γ; PUFA = polyunsaturated FA; PYY = peptide YY; SFA = saturated FA; TZD = thiazolidinedione.

example of treatment of T2D patients with a very low caloric diet (VLCD) or physical exercise. Research showed that T2D patients react differently to these treatments. When insulin resistance is mainly localized in muscles, physical exercise has a higher and faster improvement in health as compared to VLCD. In patients with insulin resistance mainly located in the liver, VLCD can normalize the glucose metabolism already within 8 weeks. However, when β-cell capacity is not sufficient enough it is known that the patient has neither benefit from VLCD nor from physical exercise [9–12]. This allows possibilities for systems interventions based on the diagnosis of decreased flexibility of specific health-related processes. In taking this concept further, towards all aspects of phenotypic flexibility, table 1 gives an example how systems diagnosis and related interventions can be created for T2D. A first beautiful example of such an approach is given in a recent publication on the reversal of cognitive decline by Bredesen [13]. This report describes a novel, comprehensive, and personalized therapeutic program that is based on the underlying pathogenesis of Alzheimer's disease, where 9 out of the 10 early Alzheimer patients displayed sustained cognitive improvement.

Need for a Timeline of the Health Trajectory

Ideally, phenotypic flexibility biomarkers develop into two dimensions. Firstly, from a single process to the complete system of flexibility ('systems flexibility biomarker', described above), and secondly, along the timeline of an individual's health trajectory, building the life story of systems flexibility, a 'biopassport'. Loss of phenotypic flexibility is a process that develops over the time span of many years. Interventions are most successful in early stages, when full reversal is possible. The storage and availability of biomarkers have been common practice in longitudinal cohorts, but the translations of its results into health care is a tediously slow process. On the other hand, (personal) health care data are collected in a fragmented (case-by-case) manner and usually not available in a structured and understandable manner for the citizen to valorize for his personal health. Since lifestyle-related health is primarily dependent on self-management and self-empowerment, it is vital that the citizen/consumer/patient has access to all relevant health data and information [14]. If biomarkers of phenotypic flexibility are the key in optimizing metabolic health, and in the prevention and treatment of metabolic diseases, they need to be measured at regular intervals. At this moment, this is neither practical nor affordable, and, moreover, most health care systems do neither focus on nor reimburse preventive diagnostics. Therefore, new diagnostic applications need to be developed which are cost-effective and minimally invasive and preferably suitable for 'do-it-yourself' applications. Developments, both in personal health portals, e.g. ITC (Information Communication Technology) for Health, and in diagnostics, e.g. 'gadgets' or dried blood spot diagnostics, are rapidly elucidating this area. The 'Nutrition Researcher Cohort' https://humanstudies.tno.nl/nrc/ [15] aims at professionalizing this movement. This biopassport is the ideal starting point for the design of both (food-based) personal health optimization and self-empowerment strategies.

Conclusion: From Products to Services

The food and nutrition market faces major challenges. The Western world suffers from too much and relatively cheap food with low nutrient, but high caloric density, mostly derived from low-cost ingredients like vegetable fat and sugars. This is a trend rapidly adapted by the developing world [16]. The food industry finds difficulty in providing scientific evidence that their products are healthy or have added health value [17]. The two key solutions here are the availability of foods with substantiated health benefits and the facilitation of personal healthy food choices. Biomarkers of phenotypic flexibility, which refocus on the assessment of health instead of disease, can help in the design and performance of science-based nutritional interventions that allow to evaluate health improvement in (apparently) healthy consumers.

A sustainable shift in eating habits towards healthier diets will not be easy to achieve. Clearly, individuals themselves are directly responsible for what they eat. However, in a complex interplay, many external agents (regulators, industrial sectors, medical professionals, the media, and social networks) influence the choices individuals make [18]. Making healthier choices is critical for the future health of our bodies and our societies. An essential part in this process is the individual's self-empowerment in making these choices, by having access to reliable information on food products and on one's personal health status through personal access to longitudinal systems flexibility diagnostics, as described above. In shaping this new reality, self-empowerment needs to be embedded in, and possibly even become the driver of, a new health care economy based on personal data ownership [14]. The development of systems diagnosis with preventive and personalized interventions may create and trigger a series of commercial service-based health industry activities in the area of diagnostics, personal food solutions, food-pharma combinations, and health advice systems, for example. Food companies may shift their product portfolio from product branding to product-service combinations (personalized products connected to a diagnostic service), food services may be integrated into a health-based personal portfolio, and ICT services will emerge based on a personal biopassport (interpretation of an individual's health data and relate this to nutrition and lifestyle advice). All of this needs to be developed based on evidence-based science and within adequate regulatory-ethical frameworks. In other words, there is some work to be done. Yet, the next generation of biomarkers of health is not only urgently needed but will also open the door to a new cost-effective model of health and health care.

Disclosure Statement

The authors declare that no financial or other conflict of interest exists in relation to the contents of the chapter.

References

1 World Economic Forum: The Global Economic Burden of Non-Communicable Diseases. Geneva, World Economic Forum, 2011.
2 van Ommen B, van der Greef J, Ordovas JM, Daniel H: Phenotypic flexibility as key factor in the human nutrition and health relationship. Genes Nutr 2014;9:423.
3 de Wit NJW, Afman LA, Mensink M, Müller M: Phenotyping the effect of diet on non-alcoholic fatty liver disease. J Hepatol 2012;57: 1370–1373.
4 van Ommen B, Keijer J, Heil SG, Kaput J: Challenging homeostasis to define biomarkers for nutrition related health. Mol Nutr Food Res 2009;53:795–804.
5 Weseler AR, Bast A: Pleiotropic-acting nutrients require integrative investigational approaches: the example of flavonoids. J Agric Food Chem 2012;60:8941–8946.
6 Bouwman J, Vogels JT, Wopereis S, et al: Visualization and identification of health space, based on personalized molecular phenotype and treatment response to relevant underlying biological processes. BMC Med Genomics 2012;5:1.
7 DeFronzo RA: Insulin resistance, lipotoxicity, type 2 diabetes and atherosclerosis: the missing links. The Claude Bernard Lecture 2009. Diabetologia 2010;53:1270–1287.
8 Abdul-Ghani MA, Tripathy D, DeFronzo RA: Contributions of β-cell dysfunction and insulin resistance to the pathogenesis of impaired glucose tolerance and impaired fasting glucose. Diabetes Care 2006;29:1130–1139.
9 Lim EL, Hollingsworth KG, Aribisala BS, et al: Reversal of type 2 diabetes: normalisation of beta cell function in association with decreased pancreas and liver triacylglycerol. Diabetologia 2011;54:2506–2514.
10 Dela F, von Linstow ME, Mikines KJ, Galbo H: Physical training may enhance beta-cell function in type 2 diabetes. Am J Physiol Endocrinol Metab 2004;287:E1024–E1031.
11 Burns N, Finucane FM, Hatunic M, et al: Early-onset type 2 diabetes in obese white subjects is characterised by a marked defect in beta cell insulin secretion, severe insulin resistance and a lack of response to aerobic exercise training. Diabetologia 2007;50:1500–1508.
12 Snel M, Gastaldelli A, Ouwens DM, et al: Effects of adding exercise to a 16-week very low-calorie diet in obese, insulin-dependent type 2 diabetes mellitus patients. J Clin Endocrinol Metab 2012;97:2512–2520.
13 Bredesen DE: Reversal of cognitive decline: a novel therapeutic program. Aging (Albany, NY) 2014;6:707–717.
14 Hafen E, Kossmann D, Brand A: Health data cooperatives – citizen empowerment. Methods Inf Med 2014;53:82–86.
15 van Ommen B: The nutrition researcher cohort: toward a new generation of nutrition research and health optimization. Genes Nutr 2013;8:343–344.
16 FAO: The State of Food Insecurity in the World 2012. Rome, FAO, 2012.
17 Katan MB: Why the European Food Safety Authority was right to reject health claims for probiotics. Benef Microbes 2012;3:85–89.
18 Vandenbroeck P, Goossens J, Clemens M: Foresight: Tackling Obesities: Future Choices – Building the Obesity System Map. London, UK Government Office for Science, 2007.
19 van Ommen B, Fairweather-Tait S, Freidig A, et al: A network biology model of micronutrient related health. Br J Nutr 2008; 99(suppl 3):S72–S80.

Baetge EE, Dhawan A, Prentice AM (eds): Next-Generation Nutritional Biomarkers to Guide Better Health Care. Nestlé Nutr Inst Workshop Ser, vol 84, pp 35–46, (DOI: 10.1159/000436950)
Nestec Ltd., Vevey/S. Karger AG., Basel, © 2016

Bioinformatics: Novel Insights from Genomic Information

Robert E.W. Hancock

Centre for Microbial Diseases and Immunity Research, University of British Columbia, Vancouver, BC, Canada

Abstract

While scientific methods have dominated research approaches in biology over the past decades, it is increasingly recognized that the complexity of biological systems must be addressed by a different approach, namely unbiased research involving the collection of large amounts of genome-wide information. To enable analysis of this information we and others are developing a variety of computational tools that allow bioinformaticists and wet laboratory biologists to extract novel patterns of data from these results and generate novel biological insights while generating new hypotheses for testing in the laboratory. There are two types of critical tools, databases to collate all information on biomolecules, especially interactions, and tools that reorganize information in a supervised (e.g. pathway analysis or gene ontology) or unsupervised (nonhierarchical clustering and network analysis) manner. Here we describe some of the tools we have developed and how we have used these to gain new ideas in the general area of infection and innate immunity/inflammation. In particular, it is illustrated how such analyses enable novel hypotheses about mechanisms associated with diseases and the mechanisms of action of immunomodulatory and other interventions, the definition of mechanism-based biomarkers/diagnostics, and prospective new interventions based on drug repurposing.

Introduction

As biology becomes more complex through the application of high-throughput approaches, it is critical to develop biologist-friendly tools to permit the integration of new information with existing knowledge. Postgenomic approaches to functional analysis have tended to adopt the big science mode employing techniques dependent on high-throughput gathering of information. Current major methods include RNA-Seq (that has supplanted microarrays) for transcriptomic analysis, genome-wide association by comprehensive sequencing, CHiP-Seq for determining the binding site of transcription factors, CRISPR (clustered regularly interspaced short palindromic repeats) methods for rapid mutagenesis, which also help to generate mutant libraries, high-throughput mass spectrometry for proteomic analysis, and metabolomic methods. This overwhelming barrage of information requires the utilization of high-end sophisticated computational tools. These tools and their use, collectively termed bioinformatics, attempt to cluster/collate functional genomic information in either a supervised or unsupervised manner, in attempt to correlate global changes with the biological events that drove those changes. Thus, bioinformatics describes the two related analytical methods whereby 'bio' describes the intent to obtain biological insights while 'informatics' defines the computational nature of analyses that are necessary due to the shear amount of information being collected. Rather than comprehensively describing the vast field of informatics I will present a personal overview of our own philosophies regarding the analysis of high-throughput transcriptomic data, with illustrations from our own research. We provided a broader perspective, with specific reference to the immune system, some time ago [1], and the reader is directed to other resources for alternative perspectives [2–5].

New bioinformatic tools enable the supervised clustering of high-throughput transcriptomic data according to gene ontologies (functional descriptions), pathways, known interactors, and transcription factor binding sites (TFBS) upstream of dysregulated genes. They also enable unsupervised clustering according to global patterns of genes which are dysregulated (termed nonhierarchical clustering and often performed by creating heat maps). In addition, they also facilitate network analysis which attempts to interrelate dysregulated genes, where the proteins expressed by these genes are interconnected by their known tendency to interact within cells. This then enables one to determine key subnetworks and hubs/bottlenecks that define information flow within these networks. The supervised clustering methods are useful because they describe the dysregulated genes in terms of known biology. However, in our opinion, unsupervised clustering and network analysis methods are far more powerful because they do not make a priori assumptions that interpret new knowledge in terms of known events in cells but rather enable the determination of emergent properties, which are basically new biological insights into the processes driving the transcriptional differences observed. The results are always framed as hypotheses rather than knowledge per se and must be eventually tested in wet laboratory experiments but provide a powerful method of discovering new biology. I will illustrate how such analyses enable novel hypotheses about mechanisms associated with diseases or environmental factors and about the mechanisms of action of immunomodulatory and other interventions, and also allow to define mechanism-based biomarkers that can be used to diagnose disease syndromes and monitor/reveal the success of existing interventions. I will also describe strategies to define prospective new interventions based on drug repurposing.

Virtually any complex biological event can be investigated ranging from genetic disease to the influence of environment, nutrition, and interventions. Our own interests lie in innate immunity, which is a coordinated system of both specialized and nonspecialized immune cells, and serves as the body's first line of defense against pathogenic organisms. Unlike the adaptive arm of the immune system, innate immunity requires no previous exposure to threats, lacks true 'memory', and is an intrinsically hard-wired response. The innate immune system is essential for human survival, yet the outcome of an overly robust and/or inappropriate immune response can paradoxically result in harmful sequelae, including almost all known diseases such as cancer, atherosclerosis, ischemic heart disease, asthma, inflammatory bowel diseases, arthritis, and vasculitis. Similarly poor (or excessive) nutrition and various environmental factors can also cause immune dysfunction [5]. Regulatory networks that govern innate immune processes provide for the dynamic homeostatic control necessary to identify pathogens, amidst the normal host flora, and mount an appropriate response while minimizing host toxicity. These pathways interface to create a single system capable of selectively amplifying and integrating signals in a coordinated manner. As a consequence of its interconnectivity, dysfunctions in an innate immune pathway can destabilize the entire system, causing human inflammatory diseases that are acutely toxic or chronically debilitating. Importantly, while having inflammation as a common factor immune dysfunction shows substantial heterogeneity. For example, Goh et al. [6] used this framework to create a human 'diseasome' network, highlighting transcriptional level similarities, and differences, between disease types and also within tissue types. A similar type of meta-analysis was carried out by Jenner and Young [7], who identified a transcriptional program of 'common host response to infection' across 35 microarray experiments that –when combined – pooled the responses of 16 types of immune effector cells to various pathogenic stimuli. However, a meta-analysis need not be as broad, as demonstrated by Hampton et al. [8] by mining the data from 4 different cystic fibrosis (CF) microarray data sets to generate novel evidence that CF cells have an intrinsic defect in MHC processing and antigen presentation.

Anti-inflammatory therapeutics often have proven clinical benefits during treatment (e.g. statins in atherosclerosis) but again a good anti-inflammatory treatment for one disease may not work for another implying substantial heterogeneity in immune/inflammatory dysfunctions. A large variety of proteins are involved in inflammation, and many of them can be affected by genetic mutations that impair or otherwise dysregulate the normal function or expression of that protein. The reason why inflammation can become chronic in some diseases is still open to debate, but might be different for each type of disease. We are thus developing strategies and tools to understand the complex nature and heterogeneity of the dysregulation of innate immunity/inflammation in humans [9–14]. We propose this will shed light on the most appropriate strategies to treat human diseases and syndromes, including inflammation, as well as provide mechanism-based biomarkers.

Systems Biology Tools

Our flagship program, InnateDB [9, 10], provides the basis for understanding biological connections in cells, according to known interactions between genetic elements (such as proteins). Collectively, these elements are termed the interactome and reflect a number of studies that demonstrate the physical, metabolic, or regulatory interactions of proteins. For example TRAF6 in humans is usually depicted as having a role in the major TLR4 to NF-κB pathway of innate immunity. However it has been experimentally documented to interact with 643 other proteins, with a further 99 predicted interactions in man (www.innatedb.com). This means that there is a massive potential for this protein to bridge multiple biological pathways and events when activated by innate immune stimuli (e.g. infection, disease, and inappropriate nutrition).

InnateDB is an open-source, publicly available database and systems biology analysis platform of all of the genes, proteins, molecular interactions, pathways, and signaling responses involved in human, mouse, and bovine innate immune responses. It is becoming an important tool in immunology as evidenced by the >6,000,000 hits from more than 55,000 visitors annually. While all known pathways (>3,500) and molecular interactions (316,000 in human) are present, the emphasis on innate immunity is achieved through the contextual review, curation, and annotation of molecular interactions and pathways involved in innate immunity. To date, the InnateDB curation team has reviewed more than 5,000 publications annotating >25,000 molecular interactions of >8,700 separate genes in rich detail, including annotation of the cell, cell line, and tissue type; the molecules involved; the species; the interaction detection method, and the publication source. By including interaction and pathway data relevant to all biological processes, a much broader perspective of innate immunity can be achieved, especially since an effective innate immune response requires the coordinated efforts of many disparate processes including the endocrine, circulatory, and nervous systems [1]. Additionally, one is able to investigate any biological signaling process of interest beyond the immune system.

InnateDB facilitates systems level analyses by enabling the integration, analysis, and visualization of user-supplied quantitative data, such as gene expression data, in the context of molecular interaction networks and pathways. This includes the statistically robust analysis of overrepresented pathways, interactomes, ontologies, TFBS, and networks. One can, for example, refine the network to show only molecular interactions between a list of differentially expressed genes (and their encoded products) or view all potential interactors regardless of whether they are differentially expressed. This can aid in the identification of important nodes that may not be regulated transcriptionally or which are expressed at an earlier or later time. Each network can be interactively visualized at the click of a button using the Cerebral plug-in for Cytoscape, which has been developed as part of the InnateDB project, and generates biologically intuitive, pathway-like layouts of networks [11]. Other visualization tools are also available. Network analysis is also available through other Cytoscape plug-ins as well as our new program NetworkAnalyst [12], which can then be easily utilized for further network analysis. Conversely, gene sets can be downloaded in many different forms for analysis using additional network analysis software (the R statistical package, for example, provides a number of graph/network algorithms).

While InnateDB is our most utilized program, we have recently built other tools. These include a platform for the meta-analysis of transcriptomic, proteomic, and metabolomic data, INMEX (www.inmex.ca [13]), an interactive customizable heat map visualization program, INVEX (www.invex.ca [14]), and the extremely fast network and hub analysis and visualization tool, NetworkAnalyst (www.networkanalyst.ca [12]) which now contains all 3 modules. In the following, we will illustrate how these tools can be applied.

Mechanistic Insights and Biomarkers

Salmonella enterica sv. Typhimurium normally causes self-limiting gastroenteritis and food poisoning in our society. In AIDS patients in Africa, it has been found to be associated with severe invasive disease, which is normally associated

Table 1. Gene ontology (functional categorization) and pathway analysis of transcriptional responses attributable to iNTS in AIDS patients

Data sets compared	DE genes	Gene ontology (genes)	Pathways	Other DE genes of interest
Acute iNTS (n = 25) cf. HIV+ (n = 14)	1,214 upregulated	Cell cycle (n = 55) DNA replication (n = 25) DNA translation (n = 18) DNA repair (n = 26) Mitosis (n = 39) Cell division (n = 36)	Viral mRNA translation Viral replication Cell cycle Nucleotide excision repair	ARG1, CEACAM 6, complement, DICER, FCGR1A, IFNγ, MPO, MSR1, NFKBIB, PI3K, REL, SIGIRR, SOCS4, SOCS7, STAP2, TIRAP
Acute 'other' infections (n = 6) cf. HIV+	1,199 upregulated	Innate immune response (n = 51) Inflammatory response (n = 24)	IL-1 signaling IL-4 signaling Complement and coagulation Atypical NF-κB	IL10RB, IL18R1, IL18RAP, IFNGR1, CCR9, FAS, CARD17

Examined were differentially expressed (DE) genes when compared to patients with just AIDS (HIV+). Lack of a classical innate immune response and increased viral replication signatures were observed in contrast to differential responses attributable to 'other' *(E. coli and S. pneumoniae)* infections. Under other DE genes of interest, those with products associated with suppression of inflammatory responses are underlined.

with another species of *Salmonella* (*S. typhi*). Invasive nontyphoidal *Salmonella* (iNTS) is associated with rapid clinical deterioration in patients with HIV infections. We determined the transcriptional responses of 25 patients with underlying HIV infections complicated with iNTS, 14 patients with HIV infections without iNTS, and 6 HIV patients complicated with other acute bacterial infections (primarily *Escherichia coli*, a close relative of *S. enterica*, and *Streptococcus pneumoniae*) [15]. Around 1,200 genes were upregulated in both groups of infected patients compared to patients with HIV without a bacterial infection (table 1). However, supervised clustering revealed that there were profound differences. While upregulated genes from patients with acute infections were described as having ontologies and pathways typical of innate immune/inflammatory responses, this was not evident in patients with iNTS and could be explained by the upregulation, in these patients, of genes with products that are associated with suppression of inflammation (NFKBIB, PI3K, REL, SIGIRR, SOCS4, SOCS7). The poor prognosis in these patients could be explained by the lack of innate immune responses to control infection but also by the obvious viral signature, which was subsequently shown to reflect increased viral load [16].

The manipulation of natural innate immunity represents a new therapeutic strategy against antibiotic-resistant infections [17]. Cationic host defense (antimicrobial) peptides, which are produced by virtually all organisms, defend against infections [18]. These peptides boost protective innate immunity while suppressing potentially harmful inflammation/sepsis. Using the principle of selective boosting of innate immunity, we have developed novel small innate defense regulator (IDR) peptides with no direct antibacterial activity, that are nevertheless able to protect against many different microbial infections and inflammatory diseases in animal models, including antibiotic-resistant infections, tuberculosis, and cerebral malaria, providing a new concept of anti-infective therapy. Given the complexity of innate immunity, we assumed that these IDR peptides would be similarly complex mechanistically. We found that the peptides entered cells and bound to intracellular receptors. To understand subsequent events, we analyzed transcriptional dysregulation in human monocytes [19, 20], including pathway overrepresentation, TFBS analysis of the upstream regions of dysregulated genes and network analysis. Thus, the prediction of pathways collectively implicated the involvement of 11 pathways, including the p38, Erk1/2, and JNK mitogen-activated protein (MAP) kinases, NF-κB, phosphatidylinositol-3-phosphate kinase, and two Src family kinases, and several of these were subsequently confirmed using biochemical methods especially involving pharmacological inhibitors and assessments of phosphorylation of pathway intermediates. Similarly, the TFBS analysis predicted that more than 15 transcription factors were involved, including NF-κB (most subunits), Creb, IRF4, AP-1, AP-2, Are, E2F1, SP1, Gre, Elk, PPAR-γ, and STAT3, and many of these were confirmed biochemically.

More globally, we obtained exciting leads regarding mechanisms using hub analysis. As mentioned, transcriptomic information can be used in conjunction with information regarding the known interactors of dysregulated proteins (contained within InnateDB) to construct a network of interacting molecules. These can then be probed mathematically to reveal proteins that interact with many other proteins that are termed hubs. Hubs are considered to be key molecules in signaling since they are highly interconnected; they are considered to receive and integrate multiple signals, and pass them on to downstream nodes. Tools exist for the extraction of key hubs from transcriptomic information such as the plug-in cytoHubba for Cytoscape (hub.iis.sinica.edu.tw/cytohubba/) and our new program NetworkAnalyst [12]. Figure 1 shows the top hubs extracted from studies where human monocytes were treated with IDR1 peptide. The top hubs (most interconnected proteins) were involved in the functioning of MAP kinases, induction of chemokines, and in anti-inflammatory pathways, particularly TGF-β and IFN-type responses. The first two correspond to known

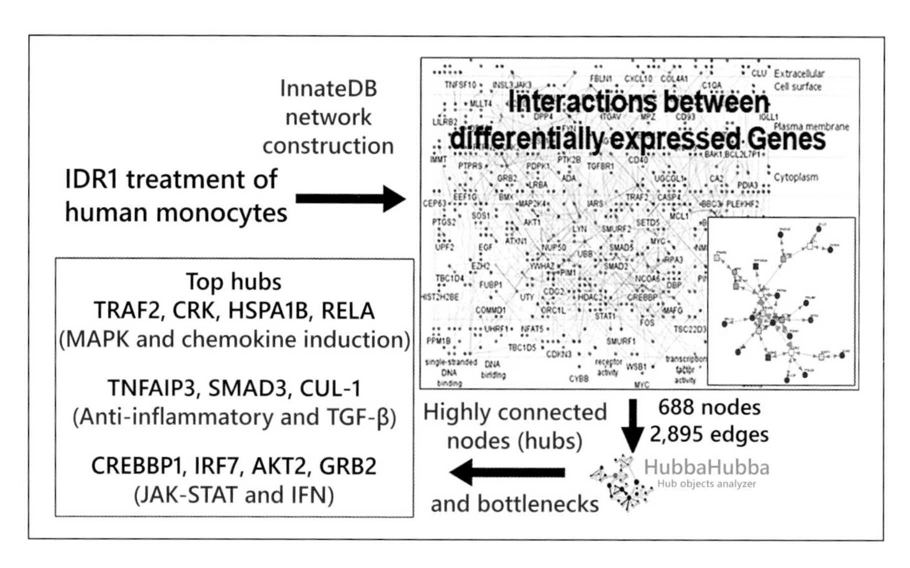

Fig. 1. Use of NetworkAnalyst to predict the key hubs in immune responses generated during IDR treatment of human monocytes. The network in the top right hand program was drawn using Cerebral and shows known (2,895) interactions (edges) between the 688 dysregulated genes (nodes: red upregulated, green downregulated). Hubs, identified using cytoHubba (Cytoscape plug-in version of HubbaHubba), are key proteins that interconnect with many other proteins and thus are predicted to be critical in the flow of information within biological signaling systems. By their nature, they are critical determinants of the mode of action and potential mechanistic-based biomarkers.

properties of the peptide, while the third is under active investigation in our laboratory with preliminary data showing that it is also involved. However, importantly, these hubs, being dysregulated and central to the network of transcriptional responses and thus biologically important, also represent excellent candidate biomarkers of the studied phenomenon and could potentially be utilized in diagnosing disease and/or testing response to treatment.

Drug Discovery and Repurposing

Another mechanistic insight was obtained in patients with CF. CF is the most common eventually fatal autosomal recessive genetic disease in our society. It is caused by mutations in the CF transmembrane regulator (CFTR), and severity and life expectancy is further influenced by modifier genes many of which impact on inflammation. Individuals with CF acquire chronic lung infections leading to hyperinflammatory lung disease, which causes progressive deterioration of lung function. Although suspected, no connectivity had been made between

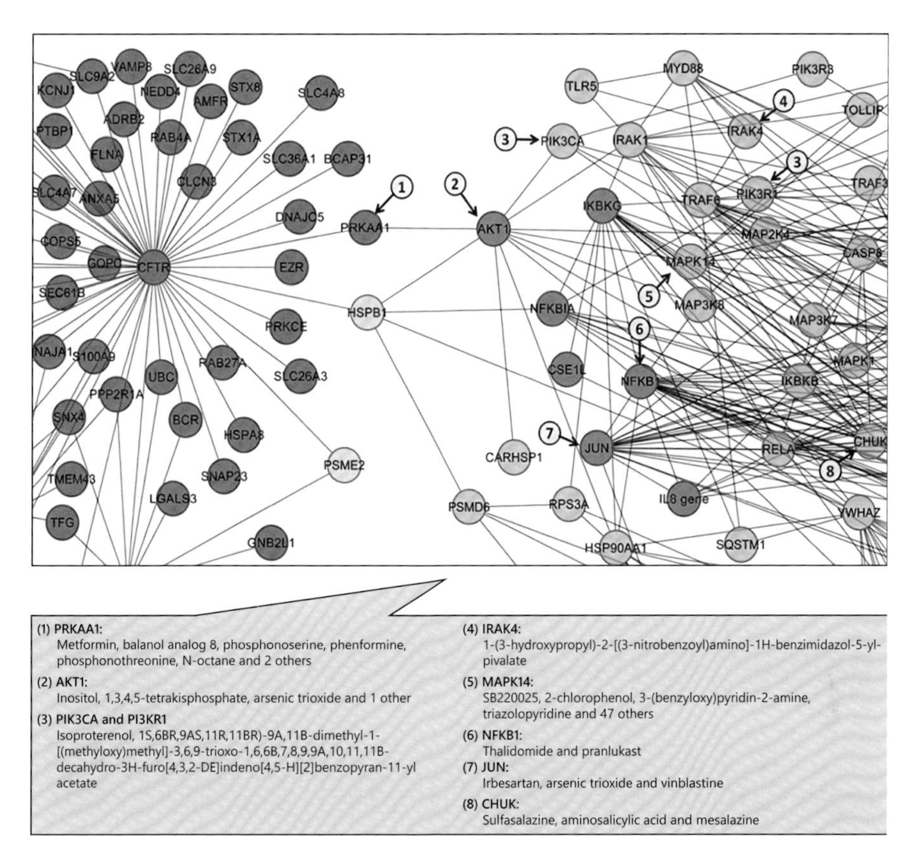

Fig. 2. Network analysis reveals potential repurposing of drugs. Analysis of transcriptomic data after challenge of CFTR mutant epithelial cells with flagellin (TLR5 agonist), cf. untreated cells, enabled the first demonstration of the linkage of the CFTR (left) and TLR5 (right) interaction networks (data from InnateDB visualized using the built-in network drawing program Cytoscape/Cerebral) and their bridging by at least 2 separate pathways (1) AKT via AMPK and (2) Hsp27. Superimposed on this are the results of the examination of these genes using the DrugBank database (see circled numbers with arrows pointing to the targeted genes), which provides all of the known pharmacophores for human gene products. Of note, metformin, a PRKAA1 (AMP kinase) activator, suppressed hyperinflammation in CF mutant cells. Reproduced from Mayer et al. [20] with permission from the American Association of Immunologists, Inc.

the CFTR status and hyperinflammation. We investigated the transcriptional responses of immortalized CFTR–/– epithelial cells compared to corrected variants with and without stimulus by flagellin, which had been shown to stimulate hyperinflammatory responses in CFTR–/– mutant cells [21]. The results, when submitted to InnateDB and a customized R-language-based assembly MetaGEX, revealed interconnectivity of the CFTR and innate immune networks through two pathways, PRKAA1 (AMP kinase)/AKT1 and HSPB1 (fig. 2). The database DrugBank [22] was then used to probe the fused network and predict inhibitors acting on the genes in this pathway that could potentially influence inflammation in CF (fig. 2). This then provided a set of repurposed drugs (many clinically approved) that could be used to influence inflammation in CF. To determine if the interconnected pathways influenced inflammation, we used the well-known approved drug metformin that activates AMP kinase and demonstrated that this suppressed inflammation by ~50%, even though AMP kinase and metformin had not previously been reported to influence inflammation.

Further, supervised mining of the genes differentially expressed in CF cells compared to corrected variants revealed the importance of stress, and, in particular, 54 genes related to the process of autophagy were dysregulated. Autophagy (also termed autophagocytosis), is a basic catabolic mechanism that involves cell degradation of unnecessary or dysfunctional cellular components through the action of lysosomes to allow the degradation and recycling of cellular components [23]. In the context of disease, autophagy is considered to be an adaptive response to stress that favors survival and in some cases is even involved in the resolution of infections, whereas in other cases it appears to promote cell death. With these bioinformatic cues about the potential involvement of autophagy in CF, we went on to show that CFTR mutant cells in fact demonstrated arrested autophagy that was not resolved, presumably converting this noninflammatory cell death mechanism into an inflammatory mechanism. We then demonstrated that the peptide IDR-1018 actually could resolve arrested autophagy and reduce inflammation.

Finally, we demonstrated, through pathway and network analysis, that there was a strong upregulation of endoplasmic reticulum stress and the unfolded protein response stress pathway, which we subsequently confirmed as a feature of CF cells [24]. This occurred through activation of IRE-1 rather than the PERK-eIF2a pathway and led us to show that salubrinal, a specific pharmacological inhibitor of the negative regulation of GADD34, upregulated this pathway and suppressed hyperinflammation.

Thus, informatic analysis not only resolved important and novel details of the nature of CF but also delivered 3 novel drugs and targets for resolving the life-threatening inflammation in CF.

The Future

These analyses only scratch the surface of what is possible using bioinformatics. I feel that the above-described examples clearly show that using unbiased experimental methods, such as whole genome transcriptome analysis in conjunction with incisive bioinformatic tools, one can go beyond the hypothesis-testing, so-called scientific method to generating fundamentally new hypotheses offering new biological insights. For us, the next frontier is (i) studying the variability of human inflammatory diseases and syndromes in order to appreciate how these various syndromes differ and how they might be individually and optimally treated and (ii) understanding the heterogeneity of any given disease. With regard to the latter, we are trying to ensure that all clinical studies collect 'metadata' that associate with each sample the parameters underlying the subject's condition, including relevant clinical data, treatments applied, and phenotypic data. With this information in hand, we are now developing new methods for multidimensional (hierarchical) clustering of metadata with transcriptional changes to develop signatures associated with each metadata parameter (age, temperature, therapies, and other defining characteristics). A prototype for this is provided by INMEX and INVEX [13, 14]. Critically, in all of our endeavors, we have tried to build informatic tools to be biologist friendly, since we believe expert biologists have the best potential to make the appropriate observations and discoveries of new relationships based on the analysis of data.

Acknowledgments

I have a Canada Research Chair which supports my salary. In addition, my bioinformatic efforts have been supported by Genome Canada, through Genome British Columbia, by the Foundation for the National Institutes of Health through their Grand Challenges in global Health programs and by the Canadian Institutes for Health Research. I particularly wish to acknowledge my collaborators Fiona Brinkman, David Lynn, and Tamara Munzer; laboratory members Jeff Xia, Erin Gill, Chris Fjell, and Jen Gardy, as well as the InnateDB curation team for their fantastic and critical contributions.

Disclosure Statement

The author declares that no financial or other conflict of interest exists in relation to the contents of the chapter.

References

1 Gardy JL, Lynn DJ, Brinkman FSL, Hancock REW: Enabling a systems biology approach to immunology: focus on innate immunity. Trends Immunol 2009;30:249–262.

2 Glynn D, Sherman BT, Hosack DA, et al: DAVID: Database for Annotation, Visualization, and Integrated Discovery. Genome Biol 2003;4:P3.

3 Zak DE, Aderem A: Systems biology of innate immunity. Immunol Rev 2009;227:264–282.

4 Kidd BA, Peters LA, Schadt EE, Dudley JT: Unifying immunology with informatics and multiscale biology. Nat Immunol 2014;15:118–127.

5 Afacan NJ, Fjell CD, Hancock REW: A systems biology approach to nutritional immunology – focus on innate immunity. Mol Aspects Med 2012;33:14–25.
6 Goh K, Cusick ME, Valle D, et al: The human disease network. Proc Natl Acad Sci U S A 2007;104:8685–8690.
7 Jenner RG, Young RA: Insights into host responses against pathogens from transcriptional profiling. Nat Rev Microbiol 2005;3: 281–294.
8 Hampton TH, Stanton BA: A novel approach to analyze gene expression data demonstrates that the DeltaF508 mutation in CFTR downregulates the antigen presentation pathway. Am J Physiol Lung Cell Mol Physiol 2010; 298:L473–L482.
9 Lynn DJ, Winsor GL, Chan C, et al: InnateDB: facilitating systems-level analyses of the mammalian innate immune response. Mol Syst Biol 2008;4:218.
10 Breuer K, Foroushani AK, Laird MR, et al: InnateDB: systems biology of innate immunity and beyond – recent updates and continuing curation. Nucleic Acids Res 2013;41: D1228–D1233.
11 Barsky A, Gardy JL, Hancock REW, Munzner T: Cerebral: a Cytoscape plugin for layout of and interaction with biological networks using subcellular localization annotation. Bioinformatics 2007;23:1040–1042.
12 Xia J, Gill E, Hancock REW: Network Analyst for statistical, visual and network-based meta-analysis of gene expression data. Nature Protocols 2015;10:823–844.
13 Xia J, Fjell CD, Mayer ML, et al: INMEX – a web-based tool for integrative meta-analysis of expression data. Nucleic Acids Res 2013; 41:W63–W70.
14 Xia J, Mayer ML, Lyle NH, et al: INVEX – a web-based tool for integrative visualization of expression data. Bioinformatics 2013;29: 3232–3234.
15 Schreiber F, Lynn DJ, Houston A, et al: The human transcriptome during non-typhoid *Salmonella* and HIV co-infection reveals attenuated NFκB-mediated inflammation and persistent cell cycle disruption. J Infect Dis 2011;204:1237–1245.
16 Preziosia MJ, Kandelb SM, Guineya DG, Brownea SH: Microbiological analysis of nontyphoidal Salmonella strains causing distinct syndromes of bacteremia or enteritis in HIV/AIDS patients in San Diego, California. J Clin Microbiol 2012;50:3598–3603.
17 Hilchie AL, Wuerth K, Hancock REW: Immune modulation by multifaceted cationic host defense (antimicrobial) peptides. Nat Chem Biol 2013;9:761–768.
18 Scott MG, Dullaghan E, Mookherjee N, et al: An anti-infective peptide that selectively modulates the innate immune response. Nat Biotechnol 2007;25:465–472.
19 Mookherjee N, Hamill P, Gardy J, et al: Systems biology evaluation of immune responses induced by human host defence peptide LL-37 in mononuclear cells. Mol Biosyst 2009;5:483–496.
20 Mayer ML, Blohmke CJ, Falsafi R, et al: Rescue of dysfunctional autophagy by IDR-1018 attenuates hyperinflammatory responses from cystic fibrosis cells. J Immunol 2013; 190:1227–1238.
21 Wishart DS, Knox C, Guo AC, et al: DrugBank: a comprehensive resource for in silico drug discovery and exploration. Nucleic Acids Res 2006;34:D668–D672.
22 Boya P, Reggiori F, Codogno P: Emerging regulation and functions of autophagy. Nat Cell Biol 2013;15:713–720.
23 Blohmke CJ, Mayer M, Fjell CD, et al: Atypical activation of the unfolded protein response in cystic fibrosis airway cells contributes to p38 MAPK-mediated innate immune responses. J Immunol 2012;189:5467–5475.
24 Blohmke CJ, Mayer M, Fjell CD, et al: A typical activation of the unfolded protein response in cystic fibrosis airway cells contributes to p38 MAPK-mediated innate immune responses. J Immunol 2012;189:5467–5475.

Baetge EE, Dhawan A, Prentice AM (eds): Next-Generation Nutritional Biomarkers to Guide Better Health Care. Nestlé Nutr Inst Workshop Ser, vol 84, pp 47–48, (DOI: 10.1159/000436952)
Nestec Ltd., Vevey/S. Karger AG., Basel, © 2016

Summary on Methodologies: Global Epidemiology

This session starts with a visionary paper from *Jim Kaput* who leads us through a brief history of classic scientific methods of the 20th century before making a convincing argument that these must be replaced, at least in part, by a new generation of thinking. Past methods have taken a reductionist approach that specifically aims to link single phenotypic outcomes (disease states) to single exposures (risk factors) and when such associations are not found, this is frequently blamed on insufficiently tight definitions of either the exposure or the outcome. Kaput argues that such reductionism flies in the face of the facts that each person is constructed from an individual genetic code overlaid by their own personal epigenetic marks, and inhabits a personal ecosystem albeit with varying degrees of overlap with others in their family, community, or country, for example. This, he argues, requires a paradigm shift to systems level thinking that now becomes tractable with the high-dimensional data produced by omic methods. The challenges are prodigious but he provides convincing examples of his vision of a next-generation systems biology approach and how this will require radical modifications to experimental design and interpretation.

Ben van Ommen and *Suzan Wopereis* start by describing physiologic and metabolic health in terms of its flexibility, its elasticity, and robustness in maintaining an optimal homeostasis in the face of different exposures (e.g. foods, nutrients, and toxins) and against different challenges (e.g. infections, exercise, and mental stress). In quantifying this phenotypic flexibility, they examine a wide range of 'stress response biomarkers' and lead us to a concept of each individual having a time-changing 'bio-passport' that will ultimately help in the design of personalized nutritional approaches that focus on health as opposed to disease.

Robert E.W. Hancock is not a nutritionist. He works on innate immune responses and was invited to share his experiences in developing advanced methodologies for interrogating very large data sets integrating genomic and metabolomic data. He shows how the complexity of biological systems requires new computational tools to extract novel biological insights. These fall into two distinct classes: supervised methods that build upon prior knowledge of metabolic pathways and gene ontologies, for example, and unsupervised (agnostic) methods that employ techniques such as nonhierarchical clustering and network analysis. Using elegant examples from his work in the field of innate immunity, he shows how fresh insights have been gained that in turn have driven new hypotheses for further testing. These methods can be extended to the development of next-generation biomarkers.

Andrew M. Prentice brings us back down to earth because, although he accepts the visions of *Kaput/van Ommen/Hancock* regarding comprehensive systems level biomarkers, it will be a long time before they can be applied to the present-day nutritional challenges faced by poor people living in third-world countries. Here, we are faced with a continual struggle to develop simplified, rugged, and affordable methods for assessing the nutrient status of populations and individuals. He alludes to international ventures to standardize such methodologies and promulgate their use in the most appropriate ways; particularly through the efforts of the BOND (Biomarkers of Nutrition and Development) initiative. One area in which he is especially optimistic about a roadmap to a next-generation point-of-care biomarker is in the use of hepcidin to indicate 'safe and ready to receive iron', a topic later covered in greater depth by Drakesmith in this volume.

Andrew M. Prentice

Baetge EE, Dhawan A, Prentice AM (eds): Next-Generation Nutritional Biomarkers to Guide Better Health Care. Nestlé Nutr Inst Workshop Ser, vol 84, pp 49–58, (DOI: 10.1159/000436954)
Nestec Ltd., Vevey/S. Karger AG., Basel, © 2016

Biomarkers in Pediatric Liver Disease

Eirini Kyrana · Emer Fitzpatrick · Anil Dhawan

Paediatric Liver, GI and Nutrition Centre, King's College Hospital, London, UK

Abstract

The chronic nature of liver diseases in children and adults merits close follow-up for disease progression and/or treatment evaluation. Disease progression involves injury to liver cells resulting in cell death, varying degrees of inflammation, steatosis depending on the insult, oxidative stress, and eventually fibrosis and cirrhosis unless the process is modified with treatment or spontaneous recovery. Inflammation, cell death, and fibrosis are the three major processes that determine the outcome of liver disease irrespective of the etiology. Markers to measure the activity or status of these parameters in a dynamic way, particularly via noninvasive methods, are urgently required. In this chapter, we summarize recent advances in the identification of biomarkers of liver diseases: biomarkers corresponding to inflammation, cell death, fibrosis, and the development of malignancy.

Introduction

Liver disease in children is rare and carries a high morbidity and mortality without liver transplantation. Advances in the diagnosis and treatment have now made it possible for children to go into adulthood. However, the chronic nature of these diseases merits closer follow-up of disease progression or treatment evaluation. Common etiological groups of liver diseases in children are inherited metabolic disorders, biliary atresia, viral hepatitis, congenital malformations, and, in the last two decades, emergence of nonalcoholic fatty liver disease (NAFLD). Disease progression involves injury to liver cells resulting in cell death, varying degrees of inflammation, steatosis depending on the insult, oxidative stress, and eventually fibrosis and cirrhosis if the process is not modified with treatment or spontaneous recovery. In viral hepatitis, the insult may be severe enough to cause acute inflammation resulting in acute liver failure (ALF), or the process may lead to chronic inflammation with/without steatosis which may cause fibrosis, cirrhosis, and sometimes hepatocellular carcinoma (HCC). Drug-induced liver injury (DILI) may also follow a process of acute inflammation culminating in ALF. In a similar way, metabolic processes may also cause acute or chronic inflammation, frequently with steatosis, and may present with ALF, or a more chronic picture of inflammation and fibrosis. Inflammation, cell death, and fibrosis are three major processes that determine the outcome of liver disease irrespective of the etiology. Markers to measure the activity or status of these parameters in a dynamic way, particularly via noninvasive methods, are urgently required.

Liver disease progression is monitored and assessed by conventional serum markers such as transaminases, bilirubin, γ-GT, alkaline phosphatase, albumin, and INR. In addition, imaging can provide insight into the liver texture or the development of liver tumors. The markers of synthetic failure (INR and albumin) are both sensitive and specific at predicting end-stage liver disease, but are of limited or no value in the early stages. Liver histology provided by liver biopsy remains the gold standard of assessing liver tissue regarding the degree and pattern of inflammation, and the presence and extent of steatosis and fibrosis. Liver biopsy, though is invasive and expensive, cannot be repeated frequently and is susceptible to sampling error. In recent years, there has, therefore, been considerable effort to develop biomarkers that could help assess the progression of liver injury noninvasively. A biomarker has been defined as a characteristic that is objectively measured and evaluated as an indicator of normal biological processes, pathogenic processes, or pharmacologic responses to a therapeutic intervention [1], and more specifically by the WHO as 'any substance, structure, or process that can be measured in the body or its products and influence or predict the incidence of outcome or disease' [1].

Biomarkers of Liver Cell Death

Toxic insults to the liver result in the death of hepatocytes mainly by apoptosis and necrosis. Apoptosis is a controlled, genetically programmed mechanism of cell death, whereas necrosis is an accidental form of cell death resulting from oxygen deprivation, mitochondrial dysfunction, and low levels of ATP. Soluble cytokeratin (CK)-18 is a major intermediate filament protein in epithelial cells and is released into the extracellular space during cell death. Measurements of CK-18 and its caspase-cleaved fragment detected by antibody M30 can monitor epithelial apoptotic cell death. Apoptotic cell death characterizes viral, cholestatic, alcoholic, and nonalcoholic fatty liver injury [2]. The fragments of CK-18 and the

M30:M65 ratio, can differentiate between apoptotic and necrotic cell death [3]. Levels of CK-18 have been shown to differentiate between patients with NAFLD and nonalcoholic steatohepatitis (NASH) [4] and have been shown to be associated with fibrosis (and not steatosis) in patients with chronic hepatitis C [5].

Members of the TNF-α family like Fas, TNF-α, and TRAIL (TNF-related apoptosis-inducing ligand) are well described inducers of hepatocyte cell death. The Fas/Fas ligand pathway has been implicated in various liver pathologies, e.g. Wilson's disease, alcoholic liver disease, ALF, chronic viral hepatitis, and NASH [3]. Higher levels of the soluble Fas receptor and the Fas ligand have been demonstrated in NASH rather than in NAFLD [6], and the soluble Fas ligand is higher in patients with ALF than in those with sepsis [7]. Levels of the soluble TRAIL receptor and TRAIL, and TNF-α have been shown to be upregulated in a variety of liver pathologies such as NAFLD, alcoholic steatohepatitis, and chronic hepatitis B and C infections [3].

HMGB1 (high-mobility group box 1) is a highly conserved, abundant, nonhistone nuclear protein expressed in almost all eukaryotic cells. Within the nucleus, HMGB1 facilitates the transcription of many genes and is passively released mainly from necrotic cells. HMGB1 is also secreted by various inflammatory cells and in this way communicating the injury to the immune system [8]. Paracetamol has also been implicated in causing ALF. HMGB1 is released from apoptotic and necrotic cells [3]. HMGB1 levels have been found to be increased in patients with acetaminophen-induced liver injury and are more sensitive than ALT levels in detecting liver injury [9].

Carbamoyl phosphate synthetase-1 is a protein found in the mitochondria mainly of the liver. It catalyzes the conversion of ammonia and bicarbonate into carbamoyl phosphate and is the first and rate-limiting step in the urea cycle. It has been proven as a potential biomarker of acute liver injury from paracetamol, Wilson's disease, or ischemia, and its levels seem to return to normal before ALT does [10].

Biomarkers of Liver Inflammation

Acute and chronic liver diseases are characterized by the presence of an inflammatory response. Chemokines are small proteins divided into 4 families (CC, CXC, CX3C, and C). Activated Kupffer cells secrete interleukin (IL)-1β and CXC chemokines which attract neutrophils which release ROS and proteases and induce hepatic necrosis [11].

Kupffer cells, injured hepatocytes, and activated hepatic stellate cells secrete CCL2 resulting in the increase in CCR2-expressing monocytes in the liver, therefore increasing the hepatic macrophage population. The liver also has a large pool of lymphocytes, and following hepatic injury, hepatic macrophages express CXCL16 which cause rapid accumulation of NKT cells (cells expressing markers of natural killer and T lymphocytes) [11].

In NAFLD, the expression of CCL2 in adipose tissue is linked with insulin resistance, adipose tissue inflammation, and hepatic steatosis. The role of Kupffer cells is central to the development of NASH, and their activation results in the recruitment of monocytes/macrophages also promoted by CCL2 [11]. There is some evidence from patients with chronic hepatitis C to suggest that complement activation reflects liver inflammation and fibrosis [12].

Biomarkers of Liver Fibrosis

Hepatic fibrosis is a result of hepatic inflammation and therefore chemokines have a role in its development. Activation of hepatic stellate cells is central to the development of fibrosis. The CCL2/CCR2 pathway is once again implicated in the development of fibrosis. CCL5 is another important chemokine pathway implicated in hepatic fibrosis as are the IFN-γ-induced chemokines CXCL9, CXCL10, and CXCL11 [11]. Other important cytokines linked with the development of hepatic fibrosis are TGF-β_1 and platelet-derived growth factor, both of which stimulate the proliferation of hepatic stellate cells. In addition, elevated serum immunoglobulin A levels have been shown to be higher in patients with NASH than in those with simple steatosis and were also an independent marker of fibrosis [13].

The identification of procollagen peptides in the serum has served as a biomarker of fibrosis. Such peptides are procollagen type-I carboxy-terminal peptide, procollagen type-III amino-terminal peptide, serum type-IV collagen, laminin, hyaluronic acid and YKL-40 [14]. Another indirect, easy-to-use marker of fibrosis, which was originally described in chronic hepatitis C, is the aspartate aminotransferase-to-platelet ratio [15]. This marker has been associated with liver fibrosis in other liver diseases, too. It was able to identify fibrosis in hepatitis C patients, particularly in those with less severe fibrosis/cirrhosis [16].

Biomarkers of Liver Cancer

Patients with chronic liver disease, particularly those with cirrhosis, are at risk of developing HCC and need close monitoring. The classic biomarker of liver malignancy is the α-fetoprotein, even though it is not specific to liver tumors, as it may be raised in chronic hepatitis and cirrhosis. Other serum biomarkers that may be used in conjunction with α-fetoprotein are des-γ-carboxy prothrombin and Golgi protein 73 [17]. Liver biopsy samples can be stained for CD10 and CD36 to diagnose HCC as well as CK-7 and CK-19 expression [17]. Transforming growth factor-β is the most important detectable growth factor in HCC; other growth factors are hepatocyte growth factor and fibroblast growth factor [17]. Elevated cytokines have also been detected, e.g. IL-6, IL-8, and IL-10 [17]. Various miRNA profiles have been described that can detect the presence of HCC on a background of chronic liver disease [18].

Biomarkers in Specific Liver Disease States

Biomarkers in Acute Liver Failure

ALF is a rare multisystem disorder in which a severe impairment in liver function occurs in association with hepatocellular necrosis in a patient with no recognized underlying chronic liver disease. In adults, encephalopathy and coagulopathy are prominent features whereas in children encephalopathy may not be present. Previously undiagnosed Wilson's disease, autoimmune hepatitis, and chronic hepatitis B or C may be included in the above definition.

ALF may resolve with conservative supportive treatment or may lead to death if liver transplant is not performed. Being able to predict which patients would recover spontaneously and which ones would need a liver transplant to survive would be invaluable to avoid unnecessary liver transplants. Various prognostic models have been developed to that effect, e.g. the KCH (King's College Hospital) criteria [19]. For acetaminophen-induced ALF, the KCH criteria include blood parameters (e.g. arterial pH, INR, and serum creatinine) as well as clinical parameters (e.g. grade of hepatic encephalopathy). For non-acetaminophen-induced ALF, the criteria include INR, age, etiology, duration of jaundice, and serum bilirubin. These criteria have been modified to include other parameters such as blood lactate and serum phosphate in order to improve the prognostic value. Another model is the MELD (model for end-stage liver disease) score, which includes serum bilirubin, creatinine, and INR [20]. CK-18 M30 has been incorporated into the ALFSG (Acute Liver Failure Study Group) index, which is superior to the KCH and MELD in predicting which patients will require liver transplantation or not [21].

Gc-globulin is a highly expressed, multifunctional and polymorphic serum protein that is synthesized in the liver. Its main function is to clear actin from the circulation during cell necrosis, and consequently to protect from disseminated intravascular coagulation resulting from the polymerization of actin. Low levels of Gc-globulin correlate with the degree of hepatic encephalopathy, and

Gc-globulin levels have been found to be lower in ALF nonsurvivors. Actin-free Gc-globulin provides the same prognostic information as the KCH criteria [22].

The protein LECT2 (leukocyte cell-derived chemotaxin-2) is a chemotactic factor for neutrophils and is identical to chondromodulin II, a growth factor for chondrocytes. It participates in the regeneration of the liver in donors and recipients after living-related transplantation. In ALF, serum LECT2 levels increased when the liver recovered [23].

Biliary carnitine (elevated medium- or long-chain acylcarnitine in bile) has been associated with a higher risk of death, lower maximum bilirubin, higher transaminases, and the presence of microvesicular steatosis in patients with acute liver injury [24]. Though being an interesting concept, measuring carnitine in bile is not practical in the clinical setting.

Biomarkers in Nonalcoholic Fatty Liver Disease/Nonalcoholic Steatohepatitis – Biomarkers of Fibrosis

The obesity epidemic has resulted in children, just as in adults, developing fatty infiltration of their liver known as NAFLD. Some of these children, though rather than having the more benign simple steatosis of NAFLD, actually have evidence of necroinflammation of the liver in the form of NASH on liver biopsy. It would be very helpful to be able to differentiate obese individuals into the ones with or without hepatic steatosis and steatosis patients into the ones with or without NASH. Studies though have shown that the important factor in prognosis and progression to cirrhosis is the extent of fibrosis present.

Markers of inflammation like ferritin and high-sensitivity C-reactive protein are associated with NASH in the context of NAFLD. High TNF-α and low adiponectin levels have also been shown to be associated. Other adipocytokines that have been implicated are visfatin, leptin, the adiponectin:leptin ratio with the HOMA-IR (homeostatic model assessment of insulin resistance), resistin, IL-6, and IL-8. Markers of cell death, as previously mentioned, have also been studied. In 201 children with biopsy-proven NAFLD, the risk of having NASH on liver biopsy increased with increasing CK-18 levels [25]. In 45 children with biopsy-proven NAFLD, CK-18 M30 levels could distinguish between controls, and children with NAFLD and children with NASH. In the same study, leptin levels could distinguish between minimal or no fibrosis and significant fibrosis [26]. Markers of oxidative stress, including lipid peroxidation, are also being investigated as potential markers of NASH. Various predictive models using a combination of clinical and laboratory parameters have also been developed to distinguish between NAFLD and NASH, e.g. the HAIR score, NashTest®, SteatoTest®, and Nash diagnostics.

In a similar way, various predictive tests have been developed and validated for the prediction of the presence of fibrosis. Examples are the pediatric NAFLD fibrosis index, which combines gender, age, body mass index, waist circumference, ALT, AST, GGT, albumin, prothrombin time, glucose, insulin, cholesterol, and triglycerides [27], or the ELF™ test [28].

Of course, hepatic steatosis can be detected by liver ultrasound particularly if it is over 30%, but ultrasound is not able to detect fibrosis. Transient elastography (Fibroscan®) has though been shown to be able to detect fibrosis and cirrhosis [29], and based on similar principles other techniques have been developed, such as acoustic radiation force impulse imaging and MR elastography.

In addition, increased expression of lumican (a keratan sulfate proteoglycan involved in collagen cross-linking and epithelial-mesenchymal transition) has been demonstrated in patients with NASH. Serum protein N-glycosylation patterns have also been shown to be different in children (and adults) with NASH, and B cells seem to have an important role in N-glycan alterations in these patients [30].

Biomarkers of Drug-Induced Liver Injury

Developing biomarkers that could help to differentiate DILI from other types of liver injury would be very important in the diagnosis of DILI. There is no pathognomonic indicator of DILI. Accurate diagnosis would be important not only for patients on drugs, but also for the development of new drugs. Ideally, a biomarker of this type would be able to discriminate between individuals susceptible to DILI, those capable of adapting to DILI, and those that can tolerate DILI [31]. We have already mentioned HMGB1 in acetaminophen toxicity, and various studies are looking into the potential role of cytokines in the development and possibly early detection of DILI [31].

More interestingly though, there are some promising prospects from the study of the role of extracellular vesicles like exosomes and microvesicles in DILI. Extracellular vesicles are released from cells into the circulation and transport signaling molecules and cellular waste, and are important cell-to-cell communicators. The content of these vesicles in mRNA and miRNA is being studied in the context of DILI with promising results and may offer the advantage of being specific to DILI. The disadvantage is currently that there is no standardized isolation method and they have a long half-life [32].

Future Trends

Micro-RNAs

No discussion about biomarkers would be complete in the current era without mentioning microRNA (miRNA). miRNAs are small noncoding RNAs (21–25 nucleotides in size) and play an important role in the regulation of gene expression. They are generated as a pre-miRNA and are exported from the nucleus. Pre-miRNA is processed into a 19- to 25-nucleotide mature form by Dicer, resulting in a double-stranded RNA molecule. Finally, a single strand is transferred into an Argonaut protein. miRNA exert biological roles when they are incorporated into the RNA inducing a silencing complex. They regulate gene expression by mRNA translation inhibition or mRNA degradation [18]. miRNA can be detected in the blood and because they are organ specific, they have a potential role as biomarkers. About 43 miRNAs have been linked to liver pathologies, but the most abundant one is miR-122. An increase in circulating miR-122 has been observed in the context of chronic hepatitis B, chronic hepatitis C, NAFLD, and acute cellular rejection after liver transplantation. Studies have been looking at the increase in combinations of miRNAs (termed signature miRNA) to differentiate between various pathologies, e.g. chronic hepatitis B in the presence or absence of HCC [18].

Conclusion

The quest for reliable, specific biomarkers which are easy to use in the clinical setting continues. The research in this area not only will help to transform the way we diagnose and monitor progression and treatment efficacy in liver disease, it will also help us gain valuable insight into the pathophysiology of the diseases.

Disclosure Statement

The authors declare that no financial or other conflict of interest exists in relation to the contents of the chapter.

References

1 Strimbu K, Tavel JA: What are biomarkers? Curr Opin HIV AIDS 2010;5:463–466.

2 Luedde T, Kaplowitz N, Schwabe RF: Cell death and cell death responses in liver disease: mechanisms and clinical relevance. Gastroenterology 2014;147:765–783.

3 Eguchi A, Wree A, Feldstein AE: Biomarkers of liver cell death. J Hepatol 2014;60:1063–1074.

4 Tsutsui M, Tanaka N, Kawakubo M, et al: Serum fragmented cytokeratin 18 levels reflect the histologic activity score of nonalcoholic fatty liver disease more accurately than serum alanine aminotransferase levels. J Clin Gastroenterol 2010;44:440–447.

5 Jazwinski AB, Thompson AJ, Clark PJ, et al: Elevated serum CK18 levels in chronic hepatitis C patients are associated with advanced fibrosis but not steatosis. J Viral Hepat 2012; 19:278–282.
6 Tamimi TI, Elgouhari HM, Alkhouri N, et al: An apoptosis panel for nonalcoholic steatohepatitis diagnosis. J Hepatol 2011;54:1224–1229.
7 Nakae H, Narita K, Endo S: Soluble Fas and soluble Fas ligand levels in patients with acute hepatic failure. J Crit Care 2001;16:59–63.
8 Lotze MT, Tracey KJ: High-mobility group box 1 protein (HMGB1): nuclear weapon in the immune arsenal. Nat Rev Immunol 2005; 5:331–342.
9 Antoine DJ, Jenkins RE, Dear JW, et al: Molecular forms of HMGB1 and keratin-18 as mechanistic biomarkers for mode of cell death and prognosis during clinical acetaminophen hepatotoxicity. J Hepatol 2012; 56:1070–1079.
10 Weerasinghe SV, Jang YJ, Fontana RJ, Omary MB: Carbamoyl phosphate synthetase-1 is a rapid turnover biomarker in mouse and human acute liver injury. Am J Physiol Gastrointest Liver Physiol 2014;307:G355–G364.
11 Marra F, Tacke F: Roles for chemokines in liver disease. Gastroenterology 2014;147:577–594.e1.
12 Vasel M, Rutz R, Bersch C, et al: Complement activation correlates with liver necrosis and fibrosis in chronic hepatitis C. Clin Immunol 2014;150:149–156.
13 McPherson S, Henderson E, Burt AD, et al: Serum immunoglobulin levels predict fibrosis in patients with non-alcoholic fatty liver disease. J Hepatol 2014;60:1055–1062.
14 Grigorescu M: Noninvasive biochemical markers of liver fibrosis. J Gastrointestin Liver Dis 2006;15:149–159.
15 Wai CT, Greenson JK, Fontana RJ, et al: A simple noninvasive index can predict both significant fibrosis and cirrhosis in patients with chronic hepatitis C. Hepatology 2003; 38:518–526.
16 Lin ZH, Xin YN, Dong QJ, et al: Performance of the aspartate aminotransferase-to-platelet ratio index for the staging of hepatitis C-related fibrosis: an updated meta-analysis. Hepatology 2011;53:726–736.
17 Mathew S, Ali A, Abdel-Hafiz H, et al: Biomarkers for virus-induced hepatocellular carcinoma (HCC). Infect Genet Evol 2014;26: 327–339.
18 Gehrau RC, Mas VR, Maluf DG: Hepatic disease biomarkers and liver transplantation: what is the potential utility of microRNAs? Expert Rev Gastroenterol Hepatol 2013;7: 157–170.
19 O'Grady JG, Alexander GJ, Hayllar KM, Williams R: Early indicators of prognosis in fulminant hepatic failure. Gastroenterology 1989;97:439–445.
20 Malinchoc M, Kamath PS, Gordon FD, et al: A model to predict poor survival in patients undergoing transjugular intrahepatic portosystemic shunts. Hepatology 2000;31:864–871.
21 Rutherford A, King LY, Hynan LS, et al: Development of an accurate index for predicting outcomes of patients with acute liver failure. Gastroenterology 2012;143:1237–1243.
22 Du WB, Pan XP, Li LJ: Prognostic models for acute liver failure. Hepatobiliary Pancreat Dis Int 2010;9:122–128.
23 Sato Y, Watanabe H, Kameyama H, et al: Serum LECT2 level as a prognostic indicator in acute liver failure. Transplant Proc 2004;36: 2359–2361.
24 Shneider BL, Rinaldo P, Emre S, et al: Abnormal concentrations of esterified carnitine in bile: a feature of pediatric acute liver failure with poor prognosis. Hepatology 2005;41: 717–721.
25 Feldstein AE, Alkhouri N, De Vito R, et al: Serum cytokeratin-18 fragment levels are useful biomarkers for nonalcoholic steatohepatitis in children. Am J Gastroenterol 2013;108:1526–1531.
26 Fitzpatrick E, Mitry RR, Quaglia A, et al: Serum levels of CK18 M30 and leptin are useful predictors of steatohepatitis and fibrosis in paediatric NAFLD. J Pediatr Gastroenterol Nutr 2010;51:500–506.
27 Nobili V, Alisi A, Vania A, et al: The pediatric NAFLD fibrosis index: a predictor of liver fibrosis in children with non-alcoholic fatty liver disease. BMC Med 2009;7:21.
28 Nobili V, Parkes J, Bottazzo G, et al: Performance of ELF serum markers in predicting fibrosis stage in pediatric non-alcoholic fatty liver disease. Gastroenterology 2009;136:160–167.

29 Friedrich-Rust M, Ong MF, Martens S, et al: Performance of transient elastography for the staging of liver fibrosis: a meta-analysis. Gastroenterology 2008;134:960–974.
30 Blomme B, Francque S, Trépo E, et al: N-glycan based biomarker distinguishing non-alcoholic steatohepatitis from steatosis independently of fibrosis. Dig Liver Dis 2012;44: 315–322.
31 Laverty HG, Antoine DJ, Benson C, et al: The potential of cytokines as safety biomarkers for drug-induced liver injury. Eur J Clin Pharmacol 2010;66:961–976.
32 Yang X, Weng Z, Mendrick DL, Shi Q: Circulating extracellular vesicles as a potential source of new biomarkers of drug-induced liver injury. Toxicol Lett 2014;225:401–406.

Baetge EE, Dhawan A, Prentice AM (eds): Next-Generation Nutritional Biomarkers to Guide Better Health Care. Nestlé Nutr Inst Workshop Ser, vol 84, pp 59–69, (DOI: 10.1159/000436955)
Nestec Ltd., Vevey/S. Karger AG., Basel, © 2016

Next-Generation Biomarkers for Iron Status

Hal Drakesmith

MRC Human Immunology Unit, Weatherall Institute of Molecular Medicine, University of Oxford, Oxford, UK

Abstract

Iron is needed for oxygen transport, muscle activity, mitochondrial function, DNA synthesis, and sensing of hypoxia. The hierarchical master determinant of dietary iron absorption and iron distribution within the body is the peptide hormone hepcidin. Hepcidin itself is regulated by a combination of signals derived from iron stores, inflammation, and erythropoietic expansion. Iron deficiency and iron deficiency anemia are common and important conditions that can be treated with iron preparations. However, other factors besides iron deficiency can cause anemia, especially inflammation, which responds poorly to iron treatment, and inherited disorders of red blood cells, which are associated with accumulation of excess pathogenic iron. Assessment of iron status is challenging, and indices such as serum ferritin, soluble transferrin receptor, and zinc protoporphyrin have specific weaknesses. Moreover, a diagnosis of iron deficiency or iron deficiency anemia is most useful if the diagnosis also leads to effective treatment. Low levels of hepcidin allow iron absorption and effective iron incorporation into red blood cells. The best 'biomarker' to guide treatment may therefore be the physiological 'determinant' of iron utilization. Iron is also important in transplantation medicine and influences clinical outcome of arterial pulmonary hypertension; here too, biomarkers including hepcidin may be useful to actively and beneficially manage iron status.

Introduction: What Is Iron for and How Do We Regulate It?

Basic cellular physiological processes require iron; hemoglobin function, oxygen sensing, generation of energy, and maintenance of genome fidelity are iron-dependent activities. Iron can easily shuttle between its ferric and ferrous valencies, and hybridize its electron orbitals to form bonds in multiple orientations, thus forming iron-sulfur complexes and heme, and allowing incorporation into enzymes. The utilization of iron into critical biochemical processes during Hadean/Archean time likely benefitted from the relatively reductive, oxygen-poor, acidic, and sulfur-rich environments prevalent in those epochs. However, since the Great Oxygenation Event ~2.3 billion years ago, iron has become poorly bioavailable due to its negligible solubility at neutral pH. Therefore, iron is now both indispensable for life but challenging to assimilate, and, due to its reactivity, toxic in excess.

In humans, requirements for iron are not constant or stable over time and vary with age (growth rate), pregnancy, and even altitude, and availability of iron is dependent on the amount and type of nutrition. Humans maintain iron homeostasis by controlling iron absorption, with iron excretion being almost unregulated [1]. An average of about 1 mg dietary Fe is taken up per day, but twenty times that amount is recycled by macrophage-mediated degradation of hemoglobin from senescent erythrocytes. Hepcidin is the 25-amino-acid peptide hormone secreted from the liver, which controls both iron recycling and absorption from the diet. Hepcidin achieves this by inhibiting the function of the iron exporter protein, ferroportin (fig. 1) [2]. Hepcidin therefore controls both the total amount of iron and its partitioning within the body. The molecular action of hepcidin and its regulation (and the relationship of hepcidin to mechanisms that maintain cellular iron homeostasis) are the subject of many reviews [1, 3, 4].

Hepcidin synthesis is controlled by three major inputs. First, accumulation of iron in serum or in the liver is sensed and leads to increased synthesis of hepcidin, which blocks iron absorption by enterocytes and macrophage recycling, returning the system to equilibrium. Genetic lesions that cause low levels of hepcidin underlie iron-overloading disorders such as hereditary hemochromatosis. Second, hepcidin synthesis is also switched on by inflammation. High levels of hepcidin cause low levels of serum iron, which may be critically protective against infection by iron-requiring blood-dwelling micro-organisms that could cause fatal sepsis. Third, because the major single requirement for iron in the body is for erythropoiesis, the loss of blood causes profound hepcidin suppression that frees up available iron stores and enhances iron absorption, to supply the bone marrow and facilitate rapid replacement of lost erythrocytes. A mediator of hepcidin suppression in this context is the recently identified erythroblast-secreted protein erythroferrone [5]. Erythropoietin stimulates erythroferrone production, and erythroferrone acts directly on the liver to suppress hepcidin. The iron overloading observed in some inherited red cell disorders (for example thalassemia intermedia) may be caused by high levels of erythroferrone secreted by erythropoietin-stimulated erythroblasts, leading to persistently suppressed

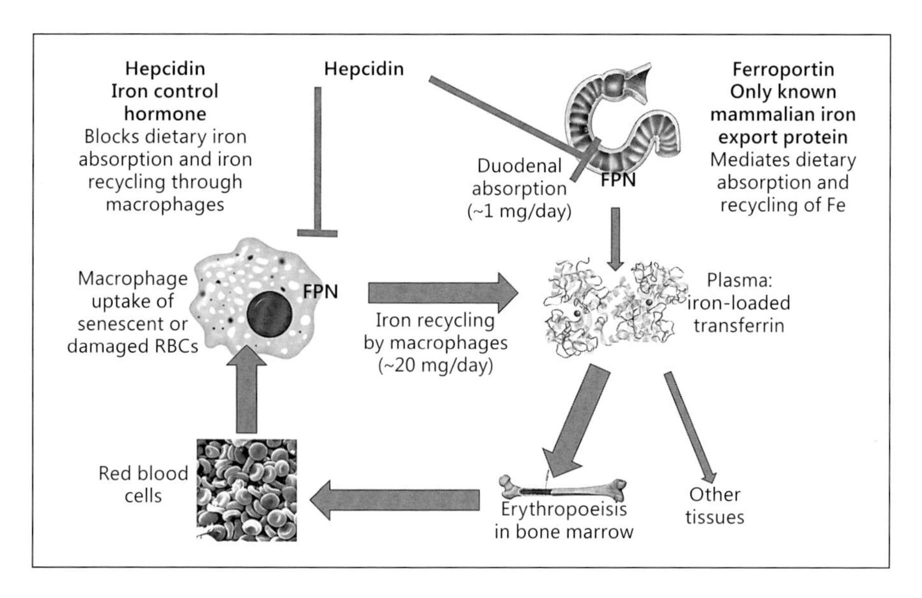

Fig. 1. Hepcidin and the iron cycle. Around 1 mg per day of heme and nonheme iron is absorbed by enterocytes in duodenal villi and transferred to serum where it is bound by transferrin – each transferrin protein can bind up to two atoms of iron. Transferrin delivers iron to tissues and cells expressing transferrin receptors; 60% of the body's transferrin receptors are in the bone marrow, where the iron is incorporated into heme in the hemoglobin of developing red blood cells (RBC). Senescent red blood cells are recognized and phagocytosed by splenic and liver macrophages. Erythrocytes are degraded, and iron is released from heme by the enzyme heme oxygenase 1. The liberated iron can either be stored within macrophages as ferritin or released back into serum. Around 20 mg of iron per day is recycled in this way. Loss of iron is not well regulated and, except in situations of blood loss including menstruation, it amounts to approximately 1 mg per day. Both the final step of iron transfer from enterocytes into plasma and the final step of iron recycling back into plasma through macrophages are mediated by a multitransmembrane protein ferroportin, the only known mammalian iron exporter. Ferroportin activity is negatively regulated by the circulating peptide hormone hepcidin that is synthesized primarily by the liver. Changes in the amount of hepcidin therefore regulate iron absorption by the intestine and iron efflux by macrophages and so determine the total amount of iron in the body and its distribution between serum and tissues.

hepcidin. Thus, hepcidin regulation incorporates signals from iron levels (liver stores and serum iron), from inflammation, and from erythropoietic demand. The signals are all believed to act at the level of the hepcidin promoter through the binding of transcription factors that are activated by the different physiological inputs, and which together control the transcription of the gene. Other levels and mediators of hepcidin regulation are known or suspected to contribute to hepcidin synthesis, but are not as well characterized as the signals discussed above and are omitted for reasons of space.

Why Do We Need to Know the Iron Status?

Because of the importance of iron for physiology, iron deficiency is associated with an array of health impairments [6], especially anemia but also suboptimal cognitive, psychomotor, and physical development, although much of the data underlying associations with developmental outcomes in infants are based on observational analyses [7]. Nevertheless, iron deficiency (usually defined using serum ferritin concentration cutoffs) is highly prevalent in infants and pregnant women in the developing world, and is thought to account for half the anemia that is present in one quarter of the world's population [8]. The socioeconomic burden of anemia is considerable, and combating anemia is a major aim of global health programs. Anemia is also highly prevalent amongst hospitalized patients in the developed world, but as discussed below, the cause of this anemia is often inflammatory rather than due to iron deficiency. Iron status has been correlated with outcome of kidney [9] and liver transplantation [10], and heart disease [11], and iron availability can influence response to hypoxic pulmonary hypertension [12], heart failure [13], as well as recovery from iron deficiency anemia. In summary, iron status is relevant for a variety of important disorders.

Why Is It Important to Guide Iron Therapy?

A variety of methods to increase iron availability are available, including iron tablets, iron-containing multiple micronutrient powders, iron fortification of foods, and iron preparations for intravenous delivery. Population-level iron supplementation is recommended in areas of high anemia prevalence, but evidence has emerged and is still accumulating that universal nontargeted oral iron treatments can have significant downsides.

First, several trials of different types of iron supplementation in different territories have indicated that iron may exacerbate the incidence and/or severity of infectious diseases, notably malaria, but also potentially respiratory infections [14, 15]. The underlying cause may be that iron deficiency is relatively protective against pathogen growth (especially in the context of malaria [16]) and that in some individuals the potential hematological benefit of iron is outweighed by the risk of developing an infection. This is a complex question and the evidence is not entirely clear or indeed consistent, but nevertheless in areas of high infectious burden and suboptimal access to health care, the infection risk of iron supplementation is a significant consideration.

A related but different issue is the effect of iron supplementation on gut microbiota. Iron excess has been shown to cause the outgrowth of potentially pathogenic gut flora at the expense of non-iron-requiring and protective lactobacilli [17]. Diarrhea is observed in populations given oral iron supplements, and this along with impaired food absorption capacity and increased intestinal inflammation, caused by imbalanced gut flora, is obviously an undesirable consequence of iron therapy. Reducing the dose of iron may theoretically in part address this problem. Intravenous iron is also available and has demonstrated beneficial profiles of hematological response, but again is associated with an increased incidence of infection [18], and its use in community populations may be compromised by financial and other considerations.

Anemia can have many causes that are not iron related, but iron therapy is very likely to be most effective as a treatment for iron deficiency anemia. An important example is the anemia of inflammation that is the most common cause of anemia in hospitalized patients. In this setting, anemia is due to the effect of inflammatory cytokines that can impair erythropoiesis and cause persistently high levels of hepcidin. The elevated hepcidin chronically downregulates the capacity for iron absorption and prevents release of iron into serum [19]. As such, anemia of inflammation is often poorly responsive to iron, and patients with this condition given iron may be more at risk of developing gastrointestinal problems due to iron being retained in the intestinal lumen. In contrast, hepcidin levels are typically low in iron deficiency, allowing utilization of dietary iron supplements. This contrast illustrates the general proposal that appropriate targeting of iron therapy is required to minimize risks and maximize benefits. Not all types of anemia are caused by iron deficiency and not all will respond to iron; iron is not harmless but is a potential cause of intestinal dysfunction and may exacerbate infections.

Is Hepcidin the Best Index to Guide Iron Therapy?

Ideally, iron should be given to individuals that are both iron deficient and able to efficiently absorb and utilize iron in the form in which it is provided. Conversely, iron could and perhaps should be withheld from individuals who are iron replete or who have inflammation. Finding biomarkers not only to diagnose iron deficiency but, equally important, also to identify the ability to absorb and utilize iron is challenging. Serum ferritin has been a key index for iron status for decades but is limited for at least two reasons. First, although low ferritin is generally associated with low iron stores and hence iron deficiency, ferritin is also an acute-phase protein and transient inflammation increases its levels, thus masking its regulation by iron. Second, ferritin is not directly influenced by erythropoietic demand for iron. In a cohort of African children of mixed iron, hemoglobin, and inflammatory status, ferritin levels were similar in anemic and nonanemic children, whereas hepcidin was lower in anemic children, likely reflecting the sensitivity of hepcidin to the iron requirement of the bone marrow [20].

Of other indices used to assess iron deficiency, soluble transferrin receptor is an indicator of erythropoietic drive, but lacks specificity for iron deficiency and is increased by hemolytic anemia and sickle cell disease; moreover, different assays give different results. Zinc protoporphyrin levels increase when there is a lack of iron to incorporate into heme, but this increase can be caused by iron deficiency or anemia of inflammation or thalassemia. Transferrin saturation is decreased by iron deficiency but also by the hypoferremic response to inflammation. Hepcidin is the only single index that incorporates sensitivity (at the transcriptional level, as described above) to iron levels, inflammation, and erythropoietic demand for iron. The log ferritin/soluble transferrin receptor index may also reflect a balance of these three inputs and appears to be a useful biomarker although again its utility may be affected by the lack of standardized soluble transferrin receptor assays as well as the need for calculation.

Serum hepcidin measurement also lacks standardization although this may be achievable to some extent by the use of adjustment factors to allow comparison between assay systems. Hepcidin possesses a further advantage over the other indices above, which may be illustrated by considering the rare autosomal recessive condition of iron-refractory iron deficiency anemia (IRIDA). Refractory iron deficiency anemia (that does not respond to oral iron) is not uncommon but is usually caused by *Helicobacter pylori* infection, autoimmune gastritis, or celiac disease [21]. However, for a small proportion of patients, a different cause has been identified. These patients have very low serum ferritin, low transferrin saturation, increased soluble transferrin receptor levels, and high zinc protoporphyrin as well as anemia, and generally no evidence of inflammation. These indices diagnose iron deficiency anemia, but this diagnosis does not lead to successful treatment. The cause of IRIDA is mutation of a gene (*TMPRSS6)* that usually restrains synthesis of hepcidin [22, 23]. In affected patients, hepcidin levels are increased, despite iron deficiency, and this increase explains both the etiology of the anemia and the resistance to oral iron therapy. By measuring the *determinant* of iron absorption and iron trafficking, more insight can be gained than by measuring other iron status indices. Although genetic IRIDA is rare, the concept illustrated is that, in general, hepcidin may most accurately report the ability to utilize iron (as we found in a study of African children of mixed-cause anemia [24]). Treatment of IRIDA is usually by intravenous iron, but even the response to this is suboptimal and short lived because iron is retained in macrophages. Antagonists of hepcidin are being developed and may have a critical role to play in IRIDA and perhaps in anemia of inflammation.

We recently demonstrated the ability of hepcidin as a single index to identify iron deficiency in a large population of African children; sex, age, wasting, and carriage of hemoglobinopathy did not significantly affect this diagnostic property of hepcidin [20]. Furthermore and importantly, we found that hepcidin measurement could also be used to categorize anemia into likely iron-responsive and non-iron-responsive types [20]. This concept is not synonymous with traditional definitions of iron deficiency, iron deficiency anemia, and anemia of inflammation, but rather hepcidin measurement may reflect the combined need for iron *and* the ability to respond to oral iron.

In individuals, anemia is likely to be due to several causes of greater or lesser contribution; tissue iron deficiency may often exist alongside moderate inflammation and erythropoietic iron demand. Attempting to understand the balance of these signals is challenging and essentially has led to altered cutoffs for serum ferritin in the presence of inflammation, and the development of the log ferritin/soluble transferrin receptor index, for example. However, the binding of different transcription factors to the hepcidin promoter also represents these same signals, and the output of the balance (increased or decreased hepcidin) appears to be the major determinant of iron absorption and utilization. *In other words, determining iron status has been a difficult and evolving empirical exercise that unconsciously attempted to measure the signals that we now know control hepcidin synthesis* (table 1). For this reason, hepcidin itself is likely to be an excellent guide for iron supplementation and could be used to target oral iron treatments more effectively.

What Are the Barriers to Implementing Screen-and-Treat Programs?

The above proposal is supported by theoretical considerations, animal model work, our work in African children, and work of others in different settings [25, 26]. However, not all analyses have led to the same conclusion [27] and it will be important to define those populations, subgroups, and disease settings where hepcidin performs best. A critical part of this is to ensure that the hepcidin assay used is able to detect low concentrations of serum hepcidin; some recent analyses have been to some extent compromised by a limit of detection that excluded many (if not most) samples in the study. Standardization of the assay will also be important in this regard.

Definitive evidence of the ability of hepcidin to guide iron supplementation is still lacking, although prospective trials are underway to assess this issue. Should they be successful, the rationale for hepcidin-guided iron supplementation (screen-and-treat programs) will be better supported, but implementation will also depend on the availability of affordable and reliable assays that work

Table 1. Equivalence of indices used to assess the iron status and known molecular regulators of hepcidin

Molecular regulators of hepcidin synthesis		Biomarkers used to assess the iron status	
signal regulating hepcidin (direction of regulation)	what is the signal derived from?	index of the iron status	what does the index represent?
BMP6/SMAD signaling (↑)	Liver iron stores	Serum ferritin	Iron stores (mostly liver)
HFE/TfR2, SMAD signaling (↑)	Iron available in blood	Transferrin saturation	Iron available in blood
IL-6/STA3 signaling (↑)	Inflammation	C-reactive protein	Inflammation
Erythroferrone (↓)	Erythropoietin signaling in bone marrow following blood loss or in thalassemia	Soluble transferrin receptor Zinc protoporphyrin	Bone marrow demand for iron Lack of iron available to incorporate into heme

Hepcidin is upregulated by iron: both by iron in the liver, which drives hepcidin transcription through bone morphogenetic protein 6 (BMP6) and SMAD signaling, and by iron in serum bound to transferrin, again via SMAD transcription factors but also requiring HFE and TfR2. Inflammation stimulates hepcidin synthesis through cytokines, especially IL-6 (although IL-22 and type-I interferon may contribute), and the transcription factor STAT3. Bone marrow demand for iron is communicated to the liver via the erythroblast-secreted hormone erythroferrone, which causes suppression of hepcidin. Commonly used indices of iron status (e.g. C-reactive protein, which is used to assess inflammation but influences the interpretation of serum ferritin concentrations) show a very close relationship with the known factors that control hepcidin, so that attempts to ascertain iron status using a combination of these indices could be viewed as measures of the balance of hepcidin regulation.

rapidly enough to inform iron treatment decisions before an individual's iron status significantly alters. One possibility may be a dipstick measurement, the development of which would not appear to present an insurmountable technological difficulty.

Are There Other Emerging Candidates for Iron Biomarkers?

The recently discovered erythroblast-derived suppressor of hepcidin, erythroferrone [5], may have some value as a biomarker, although this concept is necessarily largely theoretical at present, as the role of erythroferrone in humans has not been confirmed and assays to measure erythroferrone in human sera are not currently validated. Nevertheless, it is possible that erythroferrone could indicate the degree of erythropoietic response to recombinant erythropoietin (for example in the treatment of renal failure) and erythroferrone/hepcidin ratios could inform as to the extent of bone marrow iron demand in thalassemia, and potentially the degree of bone marrow suppression in anemia of inflammation.

What About Iron and Hepcidin in Nonhematological Conditions?

This paper has focused mainly on iron status and hepcidin in relation to anemia, but evidence from the past few years has demonstrated associations of iron and hepcidin in the contexts of kidney disease, transplantation medicine, heart failure, and response to hypoxia. There is a great deal of interest in the role of hepcidin in renal medicine, for example on whether inflammation-induced hepcidin contributes to anemia, the use of hepcidin as a marker to predict responsiveness to erythropoietin [28], as a corollary of glomerular filtration rate, and as a potential therapeutic target [29]. Hepcidin may also be of use after transplantation to monitor renal function and iron status, as iron overload is believed to be harmful following transplantation [9]. Interestingly, the reverse may be the case in liver transplantation, as high hepatic hepcidin and low transferrin receptor expression predict successful weaning from immunosuppression and tolerance to the hepatic graft [10].

Preoperative anemia is associated with higher postoperative mortality in heart surgery patients, and a major cause of this anemia appears to be functional iron deficiency (in which total iron stores may not be low, but iron is relatively unavailable for erythropoiesis) [11]. Interestingly, a recent study found that of all iron and hematopoietic indices tested, only (high) hepcidin was an independent indicator of mortality [30]. Hepcidin controls iron availability not only to the bone marrow but to cells in general – and iron is required both for muscle function and for appropriate sensing of hypoxia, both of which are likely to be important in the context of heart surgery. This finding may also be of relevance to a previous observation that intravenous iron carboxymaltose improved the functional status, symptoms, and the quality of life in patients with heart failure and iron deficiency [13]. Thus, although further prospective trials are needed, hepcidin may be of use to identify at-risk groups and guide iron therapy in the area of cardiac disease.

Concluding Remarks

Measuring iron status is difficult, as several factors need to be incorporated and balanced appropriately: iron stores, inflammation, serum iron levels, and bone marrow demand for iron. However, the regulation of hepcidin expression represents a molecular integration of these inputs, and the levels of hepcidin then determine at least in part the likely efficacy of iron therapy. The issue is important not only because of the prevalence of iron deficiency and anemia worldwide and their contribution as comorbidities to conditions such as heart disease and kidney failure, but also because iron therapy is increasingly recognized as being in some circumstances actively harmful. Therefore, targeting of iron will be important. The use of hepcidin as a guide for iron therapy is currently a justifiable concept, but a great deal more evidence in different disease conditions and settings is required, and more standardized assays may need to become available before its potential can be realized.

Acknowledgments

My laboratory is supported by the Medical Research Council UK, NIHR Oxford Biomedical Research Centre, and the Bill and Melinda Gates Foundation ('Hepcidin and Iron in Global Health', OPP1055865). I thank Sant-Rayn Pasricha and Andrew Armitage for discussion and critical feedback. Due to space constraints, important work by many colleagues could not be cited.

Disclosure Statement

The author declares that no financial or other conflict of interest exists in relation to the contents of the chapter.

References

1 Ganz T: Systemic iron homeostasis. Physiol Rev 2013;93:1721–1741.

2 Nemeth E, Tuttle MS, Powelson J, et al: Hepcidin regulates cellular iron efflux by binding to ferroportin and inducing its internalization. Science 2004;306:2090–2093.

3 Drakesmith H, Prentice AM: Hepcidin and the iron-infection axis. Science 2012;338: 768–772.

4 Hentze MW, Muckenthaler MU, Galy B, Camaschella C: Two to tango: regulation of mammalian iron metabolism. Cell 2010;142: 24–38.

5 Kautz L, Jung G, Valore EV, et al: Identification of erythroferrone as an erythroid regulator of iron metabolism. Nat Genet 2014;46: 678–684.

6 Stoltzfus RJ: Iron interventions for women and children in low-income countries. J Nutr 2011;141:756S–762S.

7 Pasricha SR, Drakesmith H, Black J, et al: Control of iron deficiency anemia in low- and middle-income countries. Blood 2013; 121:2607–2617.

8 Kassebaum NJ, Jasrasaria R, Naghavi M, et al: A systematic analysis of global anemia burden from 1990 to 2010. Blood 2014;123:615–624.

9 Schaefer B, Effenberger M, Zoller H: Iron metabolism in transplantation. Transpl Int 2014;27:1109–1117.

10 Bohne F, Martínez-Llordella M, Lozano JJ, et al: Intra-graft expression of genes involved in iron homeostasis predicts the development of operational tolerance in human liver transplantation. J Clin Invest 2012;122:368–382.

11 Cohen-Solal A, Leclercq C, Deray G, et al: Iron deficiency: an emerging therapeutic target in heart failure. Heart 2014;100:1414–1420.

12 Smith TG, Talbot NP, Privat C, et al: Effects of iron supplementation and depletion on hypoxic pulmonary hypertension: two randomized controlled trials. JAMA 2009;302: 1444–1450.

13 Anker SD, Comin Colet J, Filippatos G, et al: Ferric carboxymaltose in patients with heart failure and iron deficiency. N Engl J Med 2009;361:2436–2448.

14 Sazawal S, Black RE, Ramsan M, et al: Effects of routine prophylactic supplementation with iron and folic acid on admission to hospital and mortality in preschool children in a high malaria transmission setting: community-based, randomised, placebo-controlled trial. Lancet 2006;367:133–143.
15 Soofi S, Cousens S, Iqbal SP, et al: Effect of provision of daily zinc and iron with several micronutrients on growth and morbidity among young children in Pakistan: a cluster-randomised trial. Lancet 2013;382:29–40.
16 Gwamaka M, Kurtis JD, Sorensen BE, et al: Iron deficiency protects against severe *Plasmodium falciparum* malaria and death in young children. Clin Infect Dis 2012;54: 1137–1144.
17 Zimmermann MB, Chassard C, Rohner F, et al: The effects of iron fortification on the gut microbiota in African children: a randomized controlled trial in Cote d'Ivoire. Am J Clin Nutr 2010;92:1406–1415.
18 Litton E, Xiao J, Ho KM: Safety and efficacy of intravenous iron therapy in reducing requirement for allogeneic blood transfusion: systematic review and meta-analysis of randomised clinical trials. BMJ 2013;347:f4822.
19 Ganz T, Nemeth E: Iron sequestration and anemia of inflammation. Semin Hematol 2009;46:387–393.
20 Pasricha SR, Atkinson SH, Armitage AE, et al: Expression of the iron hormone hepcidin distinguishes different types of anemia in African children. Sci Transl Med 2014;6: 235re3.
21 Hershko C, Camaschella C: How I treat unexplained refractory iron deficiency anemia. Blood 2014;123:326–333.
22 Finberg KE, Heeney MM, Campagna DR, et al: Mutations in TMPRSS6 cause iron-refractory iron deficiency anemia (IRIDA). Nat Genet 2008;40:569–571.
23 Du X, She E, Gelbart T, et al: The serine protease TMPRSS6 is required to sense iron deficiency. Science 2008;320:1088–1092.
24 Prentice AM, Doherty CP, Abrams SA, et al: Hepcidin is the major predictor of erythrocyte iron incorporation in anemic African children. Blood 2012;119:1922–1928.
25 Bregman DB, Morris D, Koch TA, et al: Hepcidin levels predict nonresponsiveness to oral iron therapy in patients with iron deficiency anemia. Am J Hematol 2013;88:97–101.
26 Young MF, Glahn RP, Ariza-Nieto M, et al: Serum hepcidin is significantly associated with iron absorption from food and supplemental sources in healthy young women. Am J Clin Nutr 2009;89:533–538.
27 Zimmermann MB, Troesch B, Biebinger R, et al: Plasma hepcidin is a modest predictor of dietary iron bioavailability in humans, whereas oral iron loading, measured by stable-isotope appearance curves, increases plasma hepcidin. Am J Clin Nutr 2009;90:1280–1287.
28 Ashby DR, Gale DP, Busbridge M, et al: Plasma hepcidin levels are elevated but responsive to erythropoietin therapy in renal disease. Kidney Int 2009;75:976–981.
29 Ganz T, Nemeth E: Hepcidin and disorders of iron metabolism. Annu Rev Med 2011;62: 347–360.
30 Hung M, Ortmann E, Besser M, et al: A prospective observational cohort study to identify the causes of anaemia and association with outcome in cardiac surgical patients. Heart 2015;101:107–112.

Baetge EE, Dhawan A, Prentice AM (eds): Next-Generation Nutritional Biomarkers to Guide Better Health Care. Nestlé Nutr Inst Workshop Ser, vol 84, pp 71–80, (DOI: 10.1159/000436957)

The Search for Biomarkers of Long-Term Outcome after Preterm Birth

James R.C. Parkinson · Matthew J. Hyde · Neena Modi

Section of Neonatal Medicine, Department of Medicine, Imperial College London, London, UK

Abstract

Preterm birth and survival rates are rising globally, and consequently there is a growing necessity to safeguard life-long health. Epidemiological and other studies from around the world point to a higher risk of adverse adult health outcomes following preterm birth. These reports encompass morbidities in multiple domains, poorer reproductive health, and reduced longevity. The contributions of genetic inheritance, intrauterine exposures, and postnatal care practices to this altered adult phenotype are not known. Early detection is essential to implement preventive measures and to test protective antenatal and neonatal interventions to attenuate aberrant health trajectories. A satisfactory biomarker of outcome must be predictive of later functional health and ideally remain stable over the period from infancy to childhood and adult life. To date, blood pressure is the index that best fulfils these criteria. High throughput 'omic' technologies may identify biomarkers of later outcome and health risk. However, their potential can only be realized with initial investment in large, longitudinal cohort studies, which couple serial metabolomic profiling with functional health assessments across the life course.

Introduction

The number of preterm births is rising and is recognized to be a serious global health issue. The World Health Organization reports that eleven countries now have a preterm birth rate in excess of 15%; these include Pakistan and

Indonesia, with the remaining nine in sub-Saharan Africa [1]. In high-income countries, around 6–12% of births are preterm. The third trimester of development experienced by preterm infants (born below 37 weeks of gestation) differs substantially from the conditions they would have experienced in utero, exposures that may plausibly affect multiple biological pathways and organ systems. In this review, we describe current knowledge of the health of adults born preterm (excluding neurodevelopmental, neurocognitive, and neuropsychiatric disorders), discuss the suitability of existing biomarkers to predict later metabolic and cardiovascular disease, and indicate suitable approaches to advance this field.

The Adult Phenotype following Preterm Birth

In comparison to their term-born counterparts, the life course following preterm birth is marked by greater health problems at every stage. Large national cohort studies have demonstrated independent associations between preterm birth and reduced longevity, poorer reproductive outcomes, and greater morbidity in multiple domains in relation to counterparts born at full term [2, 3]. Low gestational age at birth is independently associated with increased mortality in young adulthood, and, of note, there appears to be a greater impact with increasing immaturity [2]. Preterm women, but not men, are at increased risk of having preterm offspring, and reproductive rates of both men and women are reduced by about a third to a half [3]. Being born preterm, in addition to, and independent of, being small for gestational age, is also associated with a nearly twofold increased risk of later having pregnancy complications [4]. A large number of studies have identified preterm birth as a risk factor for features of the metabolic syndrome in adulthood, notably higher blood pressure (BP), insulin resistance, and markers of cardiovascular disease. It is important to recognize that epidemiological studies are highly susceptible to confounding, and the extent to which adverse health outcomes are reflective of remediable exposures, as opposed to genetic predisposition, is unknown.

Plausible Determinants

The multitude of intrauterine and neonatal factors that may contribute to the development of a preterm phenotype are summarized in figure 1, with current evidence of outcomes in preterm compared to term infants and adults summarized in table 1.

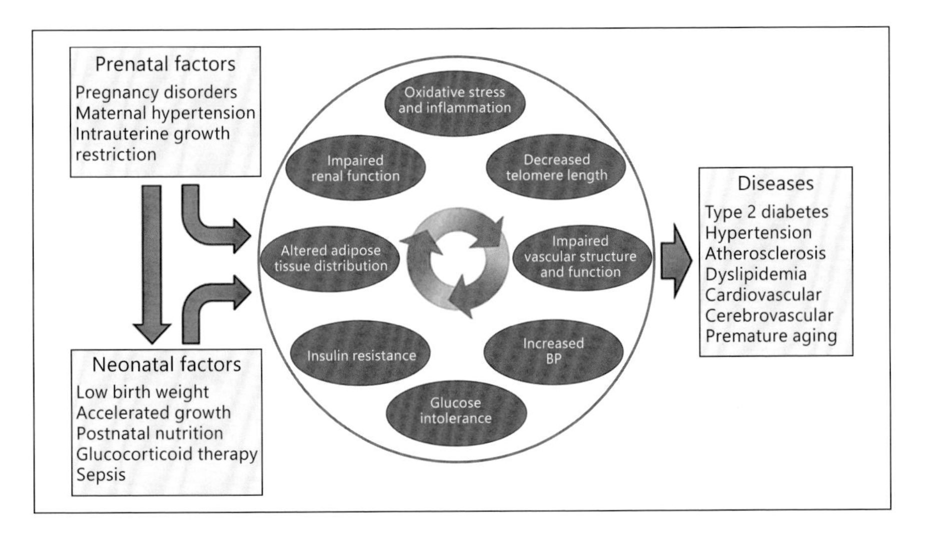

Fig. 1. The preterm phenotype.

Potential Biomarkers

Body Mass Index

Overweight and obesity are linked to a number of chronic disease states, and the risk of overweight/obese children or adolescents becoming obese adults is at least twice as high compared to their normal-weight counterparts [5]. Body mass index (BMI) might, therefore, represent an easily obtainable, reliable marker for quantifying future risk. However, avoiding overweight in childhood does not protect against the development of the metabolic syndrome should they become obese in adult life. Furthermore, obese children who go on to be normal-weight adults are not at greater risk of developing cardiovascular disease [6]. The application of anthropometric variables such as BMI to predict the development of features of the metabolic syndrome is further complicated by the fact that adults born preterm remain shorter and lighter, and have a lower BMI compared to their term born peers throughout infancy and childhood, a reduction which is attenuated during adolescence. In a meta-analysis of published studies, we found no difference in BMI between preterm- and term-born adults (pooled mean difference –0.04; 95% CI –0.33, 0.24) [7]. Taken together, these data suggest that measures of height, weight, and BMI do not effectively predict the future risk.

Body Composition and Ectopic Lipid Deposition

Bioelectrical impedance, dual X-ray absorptiometry, and other techniques used to measure percentages of fat and fat-free mass represent a potentially better marker for the cardiovascular risk than anthropometric variables [8]. However,

Table 1. Current knowledge: the phenotype of preterm compared to term-born infants and adults

	Preterm infants at term vs. term infants	Young adults born preterm vs. full term
Anthropometry		
Height, weight, and BMI	Reduced	Reduced/no difference
Waist/hip ratio	Not applicable	Increased
Metabolic syndrome		
BP	Increased	Increased
Blood glucose	Increased	No difference
Insulin sensitivity	Decreased/no difference	Decreased/no difference
Hepatic lipids	Increased	Increased
Myocellular lipids	Unknown	Increased
Internal-abdominal adipose tissue	Increased	Increased
LDL cholesterol	Unknown	Increased
Cardiovascular system		
Aortic intima-media thickness	Increased	Increased/no difference
Capillary density	Unknown	Decreased
Ventricular development	Altered	Altered
Brain structure		
Cerebrovascular tortuosity	Decreased	Unknown
White-matter microstructure	Altered	Altered
Renal tract		
Kidney size	Reduced	Reduced
Nephron number	Reduced	Reduced
Aging		
Telomere length	Decreased	Unknown
Life span	Reduced	Reduced
Reproductive health		
Pregnancy complications	Not applicable	Increased
Risk of preterm birth	Not applicable	Increased in women
Reproductive rate	Not applicable	Reduced

information on how adipose tissue depots track from infancy into adulthood is sparse, while interactions with puberty and sex further limit their application [9]. An association between lower weight of children and adolescents born preterm and reduced fat mass rather than reduced fat-free mass has been reported in several studies [10]. These conclusions were not reflected in our meta-analysis of five available studies. We identified no differences in the percent fat mass between adults born preterm and full term [7]. We have, however, identified increased intra-abdominal (visceral) adiposity and intrahepatocellular lipid levels, independently, in preterm infants [11] and adults born preterm [12], and increased intramyocellular lipid levels in adults born preterm [12]. Intra-abdominal adiposity located centrally around the organs and ectopic lipid deposits are strongly associated with inflammation, insulin resistance, and cardiovascular disease [13]. While these cross-sectional data suggest that alterations in body composition and ectopic lipid deposition persist into adult life, and thus represent potential biomarkers of outcome, larger, confirmatory longitudinal studies are required.

Insulin Sensitivity

A recent review and meta-analysis implicates preterm birth as a significant and independent risk factor for the development of both type-1 and type-2 diabetes [14]. Hyperinsulinemia and decreased skeletal muscle insulin sensitivity are key features of the metabolic syndrome and precursors to the development of type-2 diabetes. However, we identified no difference in fasting glucose or insulin between adults born preterm and full term, possibly a reflection of inconsistencies in the published literature [7]. In a systematic review examining the link between preterm birth and insulin sensitivity throughout the life course, an association was noted. However, the authors acknowledge published data are conflicting and that associations were likely to be affected by the heterogeneity of each study population and multiple confounding factors [15].

Blood Lipids

Hyperlipidemia is considered a cardinal driver of atherogenesis. Fatty streaks have been noted in the aortic wall in fetal life and peak in prevalence in infancy. Infants born to hypercholesterolemic mothers have significantly more aortic fatty streaks which persist into adolescence. Circulating levels of blood lipids track from childhood into adulthood in term-born populations, but the association is lost following adjustment for adult BMI [16]. In a meta-analysis, we noted evidence of increased LDL cholesterol in adults born preterm (pooled mean difference 0.14 mmol/l; 95% CI 0.05, 0.21 mmol/l; $p = 0.01$) but no statistically significant differences in HDL or total cholesterol [7]. A study involving the 1986 Northern Finland Birth Cohort showed that compared with term-born participants, boys born preterm had higher total cholesterol (mean 6.7%; 95% CI 0.2, 3.7%), LDL cholesterol (mean 11.7%; 95% CI 2.1, 22.3%), and apolipoprotein B (mean 12.3%; 95% CI 3.1, 22.4%); no differences were noted in girls. Differences were stronger when adjusted for maternal smoking, birth weight standard deviation score, parental education, pubertal stage, BMI, and lifestyle; similar associations were noted with gestation as a continuous variable [17]. It has been suggested that total cholesterol is a promising biomarker of long-term cardiovascular health, as maximal blood cholesterol recorded in the neonatal period has been shown to be strongly associated with greater aortic stiffness in young adults who were born preterm [18].

Cardiovascular Indices
Endothelial dysfunction is considered a precursor to the development of vascular disease. Conflicting data exist on arterial stiffness and endothelial function in children and adolescents born preterm, with a positive association found in some but not all studies [19, 20]. In a meta-analysis of the published literature, we found no significant difference between preterm and term adults in either flow-mediated dilation, intima-media thickness, or pulse wave velocity, though the number of studies was small [7]. However, sophisticated imaging techniques have recently shown that preterm birth impacts the long-term development of cardiac structure and function, with increases in myocardial mass and reductions in ventricular function [20].

Blood Pressure
Life course trajectories of BP are well established; data from a number of studies indicate that BP tracks from childhood into adulthood [21]. Higher BP in late adolescence is associated with early incidence of coronary heart disease and stroke [22]. A large number of studies have shown a clinically relevant increase of around 3–4 mm Hg in systolic BP in children and adults born preterm compared to those born at term [23]. Furthermore, a large cohort analysis has demonstrated a dose-response relationship between prematurity and BP [24]. In a meta-analysis of 13 studies, we showed that adults born preterm had clinically highly relevant increases in systolic BP (pooled mean difference 4.2 mm Hg; 95% CI 2.8, 5.7 mm Hg) and diastolic BP (pooled mean difference 2.6 mm Hg; 95% CI 1.2, 4.0 mm Hg) compared to term-born adults [7]. Ambulatory monitoring is considered a more reliable approach to assess BP as it is less affected by the anxiety response that accompanies one-off measurements. Increased BP reactivity to psychosocial stressors or 'white coat hypertension' has been observed in women born preterm, suggesting a heightened response to stress rather than a change in basal BP [25]. In the two studies that involved measurement of ambulatory BP to date, the results confirmed a difference in systolic BP in preterm compared with term-born women, but not men [7].

Metabonomics
Increasingly, the search for biomarkers has turned to the cutting-edge 'omic' technologies. These techniques generally have a high throughput, are noninvasive, and require a minimal sample volume, and thus allow screening of large numbers of potential biomarkers simultaneously. Metabonomics aims to characterize the global profile of the metabolites within a biological system, and hence provides a reflection of genomic, transcriptomic, and proteomic phenotypes in combination, i.e. an integrated profile of the biological status.

Metabonomic analysis of urine samples has identified putative biomarkers for the later development of type-1 and type-2 diabetes; for example, elevation in the glutamic acid concentration preceded the production of glutamic acid decarboxylase antibodies and overt type-1 diabetes [26], and elevation in 2-aminoadipic acid precedes the manifestation of type-2 diabetes [27]. Implementation of metabonomic technologies has revealed intriguing initial results regarding differences in urinary metabolites between preterm and term-born adults. Increased levels of metabolites were associated with inflammation and an altered microbiome, including an inverse association between urinary hippurate levels in adulthood and BP [12]. Hippurate is formed predominantly by hepatic glycine conjugation of intestinal microbial-derived benzoate produced from plant phenolics. Intestinal microbiota promote host energy recovery from dietary sources through catabolism of otherwise poorly digestible nutrients such as resistant starches, a phenomenon implicated in the development of human obesity, hypertension, and cardiovascular disease. Perturbations in the intestinal microbiome are a plausible effector pathway especially as the preterm neonate is heavily exposed to antimicrobials. Although the clinical implications of differences in the urinary metabolite profile remain uncertain, preliminary data such as these indicate a role for metabonomic technologies in the identification of biomarkers of risk in preterm populations and potential for insight into mechanisms mediating preterm birth and later health outcomes [12]. The use of metabonomics in neonatal studies has been reviewed recently [28].

Genomics and Epigenomics

It is hypothesized that high levels of oxidative stress in preterm infants reduces telomere length, explaining why they manifest signs commensurate with accelerated aging [29]. Reduced telomere length has been reported in low-birth-weight infants compared to normal-birth-weight neonates [30]. Epigenomics, the profiling of chemical alterations to the DNA without base pair sequence changes, has also been used in the search for biomarkers of later life outcome. Correlations have been demonstrated between DNA methylation in samples obtained in early life and the obesity phenotype in later life [31]. Early catch-up growth in preterm infants has also been shown to alter DNA methylation, but it remains unclear whether this methylation is the cause or the outcome of the catch-up growth [32].

Sex Differences

Sex-specific differences in biological outcomes are well recognized. An increased susceptibility to adverse outcomes in preterm men compared to preterm women has been shown in several studies in relation to adiposity and markers of inflammation [12]. Furthermore, we found the preterm-term difference in BP for women was significantly greater than the preterm-term difference in men (systolic BP, mean difference 2.9 mm Hg; 95% CI 1.1, 4.6 mm Hg; $p = 0.004$); diastolic BP (mean difference 1.6 mm Hg; 95% CI 0.3, 2.9 mm Hg; $p = 0.02$) [7]. In the 1986 Finnish Birth Cohort study, girls born preterm had a 6.7 mm Hg (95% CI 3.1, 10.2) higher systolic BP and a 3.5 mm Hg (95% CI 1.1, 5.8) higher diastolic BP than those born at term, whereas boys showed no BP differences [17]. These data suggest that certain aberrant trajectories associated with preterm birth are sex specific. This is an important consideration that should be factored in the design of clinical studies.

Conclusions

It appears clear that preterm birth is associated with adverse health outcomes across the life course. However, the extent to which adverse health outcomes reflect intrauterine or immediate postnatal experiences is unclear. This notwithstanding, it is plausible that the health care received by a preterm baby has the potential to aggravate or attenuate aberrant biological trajectories, regardless of whether these reflect genetic predisposition, the intrauterine environment, or postnatal exposures. Preterm birth may also be viewed as a natural experiment that provides the opportunity to understand the biology of human third-trimester development and to develop new approaches to slow the explosive rise in noncommunicable diseases witnessed worldwide.

A cardinal difficulty in testing new experimental approaches in preterm care, and resolving uncertainties in accepted but inadequately evidenced practices, is the need for long-term follow-up. The identification of reliable biomarkers of long-term health would immeasurably enhance the ability to test and evaluate new treatment strategies and those that are established but insufficiently evidence based. This, however, requires that long-term, longitudinal follow-up studies are conducted to identify and validate biomarkers of later health risks. The use of 'omic' technologies in the search for bio-markers of long-term outcomes of preterm birth is gathering pace. However, to date, no biomarkers with sufficiently high specificity and sensitivity for long-term outcomes of infants born preterm have been identified. This requires the comprehensive 'omic' characterization and parallel functional assessment of a large prospective cohort of preterm and term-born infants at multiple, longitudinal time points throughout their life course. The availability of large population databases holding detailed longitudinal clinical information offers the opportunity to facilitate the acquisition of information on functional outcomes. Given the increasing prevalence of preterm survivors within the total population pool globally, implementation of such a research strategy would appear to be a sound and essential investment.

Disclosure Statement

All authors declare that no financial or other conflicts of interest exist in relation to the contents of the chapter.

References

1 Beck S, Wojdyla D, Say L, et al: The worldwide incidence of preterm birth: a systematic review of maternal mortality and morbidity. Bull World Health Organ 2010;88:31–38.

2 Crump C, Sundquist K, Sundquist J, et al: Gestational age at birth and mortality in young adulthood. JAMA 2011;306:1233–1240.

3 Swamy GK, Ostbye T, Skjaerven R: Association of preterm birth with long-term survival, reproduction, and next-generation preterm birth. JAMA 2008;299:1429–1436.

4 Boivin A, Luo ZC, Audibert F, et al: Pregnancy complications among women born preterm. CMAJ 2012;184:1777–1784.

5 Singh AS, Mulder C, Twisk JW, et al: Tracking of childhood overweight into adulthood: a systematic review of the literature. Obes Rev 2008;9:474–488.

6 Lloyd LJ, Langley-Evans SC, McMullen S: Childhood obesity and risk of the adult metabolic syndrome: a systematic review. Int J Obes (Lond) 2012;36:1–11.

7 Parkinson JR, Hyde MJH, Gale C, et al: Preterm birth and the metabolic syndrome in adult life: a systematic review and meta-analysis. Pediatrics 2013;131:1240–1263.

8 Thibault R, Pichard C: The evaluation of body composition: a useful tool for clinical practice. Ann Nutr Metab 2012;60:6–16.

9 Machann J, Thamer C, Schnoedt B, et al: Age and gender related effects on adipose tissue compartments of subjects with increased risk for type 2 diabetes: a whole body MRI/MRS study. MAGMA 2005;18:128–137.

10 Fewtrell MS, Williams JE, Singhal A, et al: Early diet and peak bone mass: 20 year follow-up of a randomized trial of early diet in infants born preterm. Bone 2009;45:142–149.

11 Thomas EL, Uthaya S, Vasu V, et al: Neonatal intrahepatocellular lipid. Arch Dis Child Fetal Neonatal Ed 2008;93:F382–F383.

12 Thomas EL, Parkinson JR, Hyde MJ, et al: Aberrant adiposity and ectopic lipid deposition characterize the adult phenotype of the preterm infant. Pediatr Res 2011;70:507–512.

13 Despres JP, Lemieux I: Abdominal obesity and metabolic syndrome. Nature 2006;444:881–887.

14 Li S, Zhang M, Tian H, et al: Preterm birth and risk of type 1 and type 2 diabetes: systematic review and meta-analysis. Obes Rev 2014;15:804–811.

15 Tinnion R, Gillone J, Cheetham T, et al: Preterm birth and subsequent insulin sensitivity: a systematic review. Arch Dis Child 2014;99:362–368.

16 Freedman DS, Khan LK, Dietz WH, et al: Relationship of childhood obesity to coronary heart disease risk factors in adulthood: the Bogalusa Heart Study. Pediatrics 2001;108:712–718.

17 Sipola-Leppanen M, Vaarasmaki M, Tikanmaki M, et al: Cardiovascular risk factors in adolescents born preterm. Pediatrics 2014;134:e1072–e1081.

18 Lewandowski AJ, Lazdam M, Davis E, et al: Short-term exposure to exogenous lipids in premature infants and long-term changes in aortic and cardiac function. Arterioscler Thromb Vasc Biol 2011;31:2125–2135.

19 Skilton MR, Mikkila V, Wurtz P, et al: Fetal growth, omega-3 (n-3) fatty acids, and progression of subclinical atherosclerosis: preventing fetal origins of disease? The Cardiovascular Risk in Young Finns Study. Am J Clin Nutr 2013;97:58–65.
20 Lewandowski AJ, Leeson P: Preeclampsia, prematurity and cardiovascular health in adult life. Early Hum Dev 2014;90:725–729.
21 Wills AK, Lawlor DA, Matthews FE, et al: Life course trajectories of systolic blood pressure using longitudinal data from eight UK cohorts. PLoS Med 2011;8:e1000440.
22 Falkstedt D, Koupil I, Hemmingsson T: Blood pressure in late adolescence and early incidence of coronary heart disease and stroke in the Swedish 1969 conscription cohort. J Hypertens 2008;26:1313–1320.
23 de Jong F, Monuteaux MC, van Elburg RM, et al: Systematic review and meta-analysis of preterm birth and later systolic blood pressure. Hypertension 2012;59:226–234.
24 Johansson S, Iliadou A, Bergvall N, et al: Risk of high blood pressure among young men increases with the degree of immaturity at birth. Circulation 2005;112:3430–3436.
25 Feldt K, Raikkonen K, Eriksson JG, et al: Cardiovascular reactivity to psychological stressors in late adulthood is predicted by gestational age at birth. J Hum Hypertens 2007;21: 401–410.
26 Orešič M, Simell S, Sysi-Aho M, et al: Dysregulation of lipid and amino acid metabolism precedes islet autoimmunity in children who later progress to type 1 diabetes. J Exp Med 2008;205:2975–2984.
27 Wang TJ, Ngo D, Psychogios N, et al: 2-Aminoadipic acid is a biomarker for diabetes risk. J Clin Invest 2013;123:4309–4317.
28 Fanos V, Van den Anker J, Noto A, et al: Metabolomics in neonatology: fact or fiction? Semin Fetal Neonatal Med 2013;18:3–12.
29 Hallows SE, Regnault TRH, Betts DH: The long and short of it: the role of telomeres in fetal origins of adult disease. J Pregnancy 2012;2012:638476.
30 Entringer S, Epel ES, Kumsta R, et al: Stress exposure in intrauterine life is associated with shorter telomere length in young adulthood. Proc Natl Acad Sci U S A 2011;108:E513–E518.
31 Relton CL, Groom A, St Pourcain B, et al: DNA methylation patterns in cord blood DNA and body size in childhood. PLoS One 2012;7:e31821.
32 Groom A, Potter C, Swan DC, et al: Postnatal growth and DNA methylation are associated with differential gene expression of the TACSTD2 gene and childhood fat mass. Diabetes 2012;61:391–400.

Baetge EE, Dhawan A, Prentice AM (eds): Next-Generation Nutritional Biomarkers to Guide Better Health Care. Nestlé Nutr Inst Workshop Ser, vol 84, pp 81–88, (DOI: 10.1159/000436990)
Nestec Ltd., Vevey/S. Karger AG., Basel, © 2016

Beyond Cholesterol – New Cardiovascular Biomarkers

Harald Mangge

Clinical Institute of Medical and Chemical Laboratory Diagnostics, Medical University of Graz, and BioTechMed-Graz, Graz, Austria

Abstract

Atherosclerosis (AS) is the primary pathological result of obesity. Vulnerable AS plaques cause fatal clinical end points such as myocardial infarction and stroke. To prevent this, improvements in early diagnosis and treatment are essential. Because vulnerable AS plaques are frequently nonstenotic, they are preclinically undetectable using conventional imaging. Levels of blood lipids, C-reactive protein, and interleukin-6 are increased, but are insufficient to indicate the process of critical perpetuation before the end points present. More specific biomarkers (e.g. troponin, copetin, natriuretic peptides, growth differentiation factor-15, or soluble ST2) indicate the acute coronary syndrome or cardiac insufficiency, but not a critical destabilization of AS lesions in coronary or carotid arteries. Thus, valuable time (months to years) that could be used to treat the patient is wasted. An improved management of this dilemma may involve better detection of variations in degrees of immune inflammation in plaques by using new biomarkers in blood and/or within the lesion (molecular imaging). Macrophage and T-cell polarization, and innate and adaptive immune responses (e.g. Toll-like receptors) are involved in this critical process. New biomarkers in these mechanisms include pentraxin 3, calprotectins S100A8/S100A9, myeloperoxidase, adiponectin, interleukins, and chemokines. These proteins may also be candidates for molecular imaging using nuclear (magnetic resonance) imaging tools. Nevertheless, the main challenge remains: which asymptomatic individual should be screened? At which time interval? Intense interdisciplinary research in laboratory medicine (biomarkers), nanomedicine (nanoparticle development), and radiology (molecular imaging) will hopefully address these questions.

Introduction

Despite therapeutic advances, cardiovascular events are the leading causes of death worldwide. This is due to the increasing prevalence of atherosclerosis (AS) caused by the obesogenic lifestyle that is increasingly practiced in the so-called western, industrialized world. AS is a subacute immune-mediated inflammation of the vascular wall characterized by the infiltration of macrophages and T cells, which interact with one another and arterial wall cells [1, 2]. The ensuing chronic inflammatory process leads to the formation, progression, and rupture of vascular lesions called AS plaques [1, 3–5].

The discovery of early identification methods and techniques to follow the development of AS plaques is still an unsolved challenge, which is limited not only by the performance of blood biomarkers and imaging techniques at hand, but also by the availability of specific molecules for targeted recognition [1, 6]. Hence, AS is only diagnosed at advanced stages of the disease, either by direct assessment of the degree of stenosis or by evaluating the effect of arterial stenosis on organ perfusion [1, 7]. Nevertheless, so-called vulnerable AS plaques that contribute significantly to the end points myocardial infarction or ischemic stroke (carotid lesions) are frequently nonstenotic and thus not easily diagnosed before events occur [1].

Pathological Characteristics of the Vulnerable Atherosclerotic Lesion

Vulnerable or culprit AS plaques are vascular lesions that are prone to rupture. They are found more frequently in regions of nonuniform shear stress, e.g. around bifurcations of the carotid or coronary arteries. Vulnerable plaques are characterized by their thin fibrous cap, activated endothelium, strong infiltration of macrophages, large lipid core, immune activation, and increased production of proinflammatory mediators (cytokines, chemokines, and metalloproteinases). Adventitial neovascularization, an imbalance between clotting and bleeding, and less calcification are further features. The true role of calcification remains a matter of debate because sui generis stable calcified AS lesions may also mechanically destabilize highly inflamed plaques if they are located near these lesions.

Mechanisms Promoting Vulnerability

Inflammation

The Monocyte/Macrophage System

Immune-mediated inflammation is a key contributor to AS, and monocytes/macrophages are principal offenders. In recent years, a complex taxonomy of different macrophage subtypes (M1–M4) has been established [8]. All these subtypes contribute to the dynamic processes occurring in AS plaques. Ly6C^{high} M1 macrophages produce proinflammatory TNF-α, IL-1, IL-6, and nitric oxide; Ly6C^{low} M2 macrophages secrete anti-inflammatory IL-10. After lipid loading, both M1 and

M2 macrophages can develop into foam cells, which usually undergo apoptosis and, thus, create a necrotic center in the AS plaque. Induced by free hemoglobin/haptoglobin, M2 macrophages can develop into so-called M(Hb)-type macrophages, which produce more anti-inflammatory cytokines, show a decreased lipid uptake, an increased cholesterol efflux, and a less foamy nature. Thus, bleeding processes within plaques do not necessarily lead to destabilization. They may also alter the macrophage response to a more stabilizing one over time if an acute event does not interrupt this process [9]. Hence, complex interactions of influential factors with these macrophage types significantly affect the direction of pathological development: stabilization → destabilization → exacerbation, or chronification, respectively. Cytokines/chemokines (GM-CSF, M-CSF, LPS/IFN-γ, IL-4/IL13, and IL-10), immune complexes, metalloproteinases, and lipid peroxidation are some of the most important influential factors [10].

Adaptive Immune Responses (T-Helper Cells and B Cells)

After endothelial injury, low-density lipoprotein reaches the intima of the vessel wall and undergoes enzymatic/oxidative modifications. This modified lipoprotein is ingested by macrophages, where it supports foam cell formation and the production of proinflammatory TNF-α, IL-1β, monocyte chemoattractant-1, leukotriene B_4, and proteolytic matrix metalloproteinases (MMP). In this way, endothelial cells are stimulated to overexpress adhesion molecules like VCAM-1 and ICAM-1. Effector cells from adaptive immune responses [preferentially T-helper (Th)1/Th17 lymphocytes] accumulate in the subendothelial space around lipoproteins [11]. Macrophages and dendritic cells present various antigens (e.g. Apo B 100) through MHC class II molecules to $CD4^+$ Th cells. This mainly activates $CD4^+$ T cells of the Th1 subtype, which releases the proatherogenic cytokines IFN-γ and TNF-α. Regulatory T cells counterbalance this activation, at least in part, by the local release of TGF-β and IL-10. If this process does not subside, extralesional amplification loops come into play. Antigen-loaded dendritic cells and soluble antigens are transported via lymphatic vessels to draining lymph nodes and/or the spleen. There, naïve T cells develop into effector T cells and reenter the bloodstream. When these cells reach the AS lesion, they augment B cells such that they develop antibody-specific responses. Thus, the stability of the lesion declines, procoagulant factors are expressed, and plaque rupture and thrombosis become more probable. The gut microbiome may also substantially contribute to the extralesional amplification processes [12]. For example, the synthesis of TMAO (trimethylamine-N-oxide) from dietary phosphatidylcholine by intestinal microbiota appears to play an important role. Increased blood and urine TMAO levels have been shown to correlate positively with an increased risk of incidental cardiovascular events [13].

Intercurrent infections (e.g. common colds and influenza) may also destabilize the AS process via T-cell activation loops. Thus, Th1-cell-derived IFN-γ can increase the macrophage MMP-9/TIMP-1 production ratio. More MMP-9 and less TIMP-1 can destabilize the lesion by increased proteolysis.

The Innate Immune Response (Toll-Like Receptors)

The Toll-like receptors TLR2 [14], TLR4 [15], and TLR7 [16] are centrally involved in AS. TLR7 deficiency was shown to accelerate AS in ApoE–/– mice and to promote more vulnerable plaque phenotypes with increased lesion size, fewer smooth muscle cells, and a lower collagen content, and allowed stronger infiltration of macrophages and larger lipid deposits [16]. Furthermore, increased TLR7 expression in human carotid plaques stimulated the expression of genes associated with a more stable plaque phenotype. These included the M2 macrophage markers IL-10, IL-1RN, CD163, CLEC4A, CLEC7A, MSR1, CD36, MS4A4A, CLEC10A, CLEC13A/CD302, and CD209. In addition, genes associated with thrombosis, such as CD40L, TF/CD142, PF4/CXCL4, vWF, GPIbα/CD42b, GPIbβ/42c, GPIIb/CD41, GPIIIa/CD61, GPV/CD42d, and GPIX/CD42a are downregulated [16].

Calprotectins and Danger-Associated Molecular Patterns

The danger-associated molecular pattern proteins S100A8 and S100A9, which belong to the S100 calgranulin family, are increased by the traditional cardiovascular risk factors (smoking, obesity, hyperglycemia, and dyslipidemia). These proteins are endogenous ligands of TLR4 and the receptor for advanced glycation end products. In humans, S100A8 and S100A9 levels have been shown to correlate with the extent of coronary and carotid AS and, most importantly, with a vulnerable plaque phenotype [15]. Furthermore, Erbel et al. [17] recently found MMP7(+)S100A8(+)CD68(+) M4 macrophages in coronary artery plaque tissue of humans afflicted with increased AS instability indexes. Hence, these calprotectins represent an interesting new molecular biomarker for assessing plaque vulnerability, since they reflect the pathological process in a highly specific way.

Mast Cells

The number of mast cells and their tryptase content in human carotid AS plaques and neovascularization were positively associated with future cardiovascular events [18].

Pentraxin 3, Myeloperoxidase, and Adiponectin

In contrast to C-reactive protein, pentraxin 3, also a member of the C-reactive protein family, was not expressed in the liver, but using immunochemical analysis it was detected in the basement membrane, endothelial cells and perivascular cells of carotid endarterectomy specimens from culprit AS lesions and emboli captured by distal protection devices [19]. This fact suggests that pentraxin 3 is potentially a specific marker which can be used for molecular imaging of plaque vulnerability.

Furthermore, myeloperoxidase (MPO) was found to be positively associated with carotid plaque inflammation. Higher baseline MPO values indicated a higher baseline carotid target-to-background ratio of the most diseased segment. This relationship was even observed during a 3-month follow-up [20].

In addition, peripheral blood adiponectin levels were found to be systemically lower in patients with vulnerable, as compared to patients with stable, AS plaques [19]. We investigated the role of adiponectin and its subfractions in patients with obesity-associated pre-AS lesions [21–25], and the potential of fluorescence-labeled globular adiponectin (gAd) and the full-length form of adiponectin (fAd) subfractions (fAd) to bind to AS lesions in ApoE-deficient mice [2, 26, 27]. We found only a low binding efficiency of fAd, but an inflammation-mediated strong accumulation of gAd in the fibrous cap of AS plaques [26, 27]. Therefore, gAd may be a promising target sequence for the molecular imaging of AS lesions [26]. Furthermore, we developed nanoconstructs between gAd and PEGylated stealth liposomes [27], which can deliver a high number of signal-emitting molecules to AS lesions [27]. Other nanoconstructs between gAd and protamine-oligonucleotide nanoparticles, called proticles [28–31], displayed a particular affinity for monocytes and macrophages, which may be of interest for sequential AS plaque targeting. Thus, our results indicate the potential of gAd-targeted nanoparticles for the molecular imaging of AS.

Balance between Clotting and Bleeding

Intraplaque Hemorrhage

Undoubtedly, intraplaque bleeding plays an important role in the AS destabilization process, which leads to clinical end points. The hemoglobin/haptoglobin scavenger receptor (CD163), IL-10, HO-1 (hemoxygenase 1), ferritin, and 4-hydroxy-2-nonenal (a major product of lipid peroxidation) were found to be strongly expressed in culprit lesions of patients with unstable angina pectoris [32]. Therefore, extracellular hemoglobin from intraplaque hemorrhage appears to induce oxidative tissue damage due to heme iron and the subsequent production of reactive oxygen species. The clearance mediated by macrophage hemoglobin scavenger receptors may alter the AS process to a more aggressive one. Thus, IL-10, iron content, and HO-1 activity are markers of intraplaque bleeding, which is associated with plaque destabilization [32].

Blood Cellular Microparticles

Microparticles (MPs) are anucleoid plasma membrane fragments with a size of 50 nm to 1 μm. They consist of oxidized phospholipids and specific proteins from the cells from which they originate and are induced by pathological processes. For example, endothelial MPs (EMPs) can be induced by shear stress, angiotensin II, TNF-α, or thrombin. MPs are not 'cell debris'! Moreover, they exert endocrine and paracrine effector functions, which may play an important role in plaque vulnerability. Their levels peak during vascular remodeling when the activation of the inflammasome is greatest, and can be considered a biologic danger signal in the organism. Furthermore, they indicate prothrombogenic activity [33]. EMP levels were found to be significantly elevated in patients with carotid artery disease compared to controls [33]. Nevertheless, although research to explore the use of EMPs as biomarkers of disease has progressed substantially over the past few years, further work is still required to develop well-standardized methods for their analysis and quantification. Validation of normal EMP ranges and a specific set of EMP markers for diagnostic testing remain to be established. When a methodology with higher precision is available, analysis of EMP levels and content can be combined with other markers of AS, including inflammatory, lipid, angiogenic, and metabolic profiling. These multiplex assays may help current clinical practitioners to rank patients according to their risk of carotid disease and stroke. Thus, patients who are still asymptomatic, but who have potentially dangerous vulnerable plaque lesions, may be identified and preclinically stabilized, e.g. by preventive carotid endarterectomy [33].

Conclusions

To diagnose a vulnerable plaque phenotype well before fatal clinical end points such as myocardial infarction and stroke occur is one of the most important challenges facing personalized medicine. Here, we have shown that local and systemic biomarkers can significantly assist practitioners in improving the diagnosis and treatment for these deadly vascular lesions. Promising candidates found in the peripheral blood are the calprotectins S100A8 and S100A9, pentraxin 3, and MPO. EMPs may also be important in this context. S100A8 and S100A9 probably have a high potential, because they are centrally involved in the fine-tuning and amplification of innate immune responses through endogenous danger-associated molecular patterns. Nevertheless, it is important for therapeutic approaches that more blockers are approved for clinical testing.

Although all herein discussed biomarkers may be interesting for a therapeutic intervention, the best candidate(s) for a blocking strategy remain to be found out. Undoubtly, it is important to focuse on the immune-inflammatory pathway and to integrate aspects of a disturbed balance between clotting and bleeding usually seen in vulnerable AS lesions.

Furthermore, the following main challenges remain unsolved:

(i) To catch the vulnerable lesion at the right time in the right 'patient', and

(ii) To act in a specific and effective manner without inducing side effects, because the 'patient' may still feel healthy at the time of the successful prediagnosis.

Further interdisciplinary cooperation between laboratory medicine (biomarker research), nanotechnology (nanocarriers for use as contrast agents and/or drugs), and radiology (molecular imaging) will pave the way for future success.

Disclosure Statement

This author has nothing to disclose.

References

1 Mangge H, Almer G, Stelzer I, et al: Laboratory medicine for molecular imaging of atherosclerosis. Clin Chim Acta 2014;437C:19–24.
2 Mangge H, Almer G, Truschnig-Wilders M, et al: Inflammation, adiponectin, obesity and cardiovascular risk. Curr Med Chem 2010;17:4511–4520.
3 Libby P, Okamoto Y, Rocha VZ, Folco E: Inflammation in atherosclerosis: transition from theory to practice. Circ J 2010;74:213–220.
4 Mangge H, Becker K, Fuchs D, Gostner JM: Antioxidants, inflammation and cardiovascular disease. World J Cardiol 2014;6:462–477.
5 Mangge H, Stelzer I, Reininghaus EZ, et al: Disturbed tryptophan metabolism in cardiovascular disease. Curr Med Chem 2014;21:1931–1937.
6 Libby P, Nahrendorf M, Weissleder R: Molecular imaging of atherosclerosis: a progress report. Tex Heart Inst J 2010;37:324–327.
7 Sanz J, Fayad ZA: Imaging of atherosclerotic cardiovascular disease. Nature 2008;451:953–957.
8 Fenyo IM, Gafencu AV: The involvement of the monocytes/macrophages in chronic inflammation associated with atherosclerosis. Immunobiology 2013;218:1376–1384.
9 Finn AV, Nakano M, Polavarapu R, et al: Hemoglobin directs macrophage differentiation and prevents foam cell formation in human atherosclerotic plaques. J Am Coll Cardiol 2012;59:166–177.
10 Gleissner CA: Macrophage phenotype modulation by CXCL4 in atherosclerosis. Front Physiol 2012;3:1.
11 Ait-Oufella H, Taleb S, Mallat Z, Tedgui A: Cytokine network and T cell immunity in atherosclerosis. Semin Immunopathol 2009;31:23–33.
12 Ketelhuth DF, Hansson GK: Cellular immunity, low-density lipoprotein and atherosclerosis: break of tolerance in the artery wall. Thromb Haemost 2011;106:779–786.

13 Tang WH, Wang Z, Levison BS, et al: Intestinal microbial metabolism of phosphatidylcholine and cardiovascular risk. N Engl J Med 2013;368:1575–1584.
14 Bertocchi C, Traunwieser M, Dorler J, et al: Atorvastatin inhibits functional expression of proatherogenic TLR2 in arterial endothelial cells. Cell Physiol Biochem 2011;28:625–630.
15 Schiopu A, Cotoi OS: S100A8 and S100A9: DAMPs at the crossroads between innate immunity, traditional risk factors, and cardiovascular disease. Mediators Inflamm 2013; 2013:828354.
16 Salagianni M, Galani IE, Lundberg AM, et al: Toll-like receptor 7 protects from atherosclerosis by constraining 'inflammatory' macrophage activation. Circulation 2012;126:952–962.
17 Erbel C, Wolf A, Lasitschka F, et al: Prevalence of M4 macrophages within human coronary atherosclerotic plaques is associated with features of plaque instability. Int J Cardiol 2015;186:219–225.
18 Willems S, Vink A, Bot I, et al: Mast cells in human carotid atherosclerotic plaques are associated with intraplaque microvessel density and the occurrence of future cardiovascular events. Eur Heart J 2013;34:3699–3706.
19 Shindo A, Tanemura H, Yata K, et al: Inflammatory biomarkers in atherosclerosis: pentraxin 3 can become a novel marker of plaque vulnerability. PLoS One 2014;9:e100045.
20 Duivenvoorden R, Mani V, Woodward M, et al: Relationship of serum inflammatory biomarkers with plaque inflammation assessed by FDG PET/CT: the dal-PLAQUE study. JACC Cardiovascular Imaging 2013;6:1087–1094.
21 Mangge H, Almer G, Haj-Yahya S, et al: Preatherosclerosis and adiponectin subfractions in obese adolescents. Obesity 2008;16: 2578–2584.
22 Mangge H, Almer G, Haj-Yahya S, et al: Nuchal thickness of subcutaneous adipose tissue is tightly associated with an increased LMW/ total adiponectin ratio in obese juveniles. Atherosclerosis 2009;203:277–283.
23 Pilz S, Mangge H, Wellnitz B, et al: Adiponectin and mortality in patients undergoing coronary angiography. J Clin Endocrinol Metab 2006;91:4277–4286.
24 Pilz S, Maerz W, Weihrauch G, et al: Adiponectin serum concentrations in men with coronary artery disease: the LUdwigshafen RIsk and Cardiovascular Health (LURIC) study. Clin Chim Acta 2006;364:251–255.
25 Pilz S, Horejsi R, Moller R, et al: Early atherosclerosis in obese juveniles is associated with low serum levels of adiponectin. J Clin Endocrinol Metab 2005;90:4792–4796.
26 Almer G, Saba-Lepek M, Haj-Yahya S, et al: Globular domain of adiponectin: promising target molecule for detection of atherosclerotic lesions. Biologics 2011;5:95–105.
27 Almer G, Wernig K, Saba-Lepek M, et al: Adiponectin-coated nanoparticles for enhanced imaging of atherosclerotic plaques. Int J Nanomedicine 2011;6:1279–1290.
28 Junghans M, Kreuter J, Zimmer A: Antisense delivery using protamine-oligonucleotide particles. Nucleic Acids Res 2000;28:E45.
29 Lochmann D, Weyermann J, Georgens C, et al: Albumin-protamine-oligonucleotide nanoparticles as a new antisense delivery system. Part 1: physicochemical characterization. Eur J Pharm Biopharm 2005;59:419–429.
30 Weyermann J, Lochmann D, Georgens C, Zimmer A: Albumin-protamine-oligonucleotide-nanoparticles as a new antisense delivery system. Part 2: cellular uptake and effect. Eur J Pharm Biopharm 2005;59:431–438.
31 Wernig K, Griesbacher M, Andreae F, et al: Depot formulation of vasoactive intestinal peptide by protamine-based biodegradable nanoparticles. J Control Release 2008;130: 192–198.
32 Yunoki K, Inoue T, Sugioka K, et al: Association between hemoglobin scavenger receptor and heme oxygenase-1-related anti-inflammatory mediators in human coronary stable and unstable plaques. Hum Pathol 2013;44: 2256–2265.
33 Schiro A, Wilkinson FL, Weston R, et al: Endothelial microparticles as conveyors of information in atherosclerotic disease. Atherosclerosis 2014;234:295–302.

Baetge EE, Dhawan A, Prentice AM (eds): Next-Generation Nutritional Biomarkers to Guide Better Health Care. Nestlé Nutr Inst Workshop Ser, vol 84, pp 89–90, (DOI: 10.1159/000436991)
Nestec Ltd., Vevey/S. Karger AG., Basel, © 2016

Summary on Applications/End Users

Biomarkers for the diagnosis, prognosis, and treatment monitoring have now become an integral part of the modern management of chronic diseases in humans. The liver is the only solid organ that has an inherent regenerative potential depending on the severity and nature of the injury. Chronicity is a common natural history leading to liver failure in most of the untreated liver diseases. Liver injury leads to cell death followed by fibrosis and complications such as impaired synthesis and detoxification and the sequel of portal hypertension. Biomarkers in various body fluids continue to be evaluated and implemented in clinical practice. Standard tests of liver function or dysfunction, namely albumin, prothrombin time, and liver enzymes, are the most commonly used but suffer from being nonspecific. Apart from repeated imaging, liver biopsy continues to be the gold standard to assess the severity of liver disease, which has the downsides of being invasive and expensive as well as difficult to repeat serially. Great progress is being made in assessing liver cell damage and its mechanism of progression (apoptotic or necrotic) by serum biomarkers such as cytokeratin (CK)-18 and its fractions. The fragments of CK-18 and the M30:M65 ratio can differentiate between apoptotic and necrotic cell death. Levels of CK-18 have been shown to differentiate between patients with nonalcoholic fatty liver disease and nonalcoholic steatohepatitis, and have been shown to be associated with fibrosis (and not steatosis) in patients with chronic hepatitis C. Inflammation is usually an essential component of acute and chronic liver diseases. Mediators of inflammation, e.g. chemokines, are small proteins classified into four families (CC, CXC, CX3C, and C). Activated Kupffer cells secrete interleukin-1β and CXC chemokines, which attract neutrophils that release reactive oxygen species and proteases, and induce hepatic necrosis. Some of these inflammatory mediators are being increasingly used as prognostic biomarkers in the management of liver disease. Fibrosis is the sequel of liver cell death. Activation of hepatic stellate cells is central in the development of fibrosis. The CCL2/CCR2 pathway is once again implicated in the development of fibrosis. CCL5 is another important chemokine involved in hepatic fibrosis as well as the IFN-γ induced chemokines CXCL9, CXCL10, and CXCL. Other important cytokines linked with the development of hepatic fibrosis are TGF-β_1 and PDGF, both of which stimulate the proliferation of hepatic stellate cells. Identification of procollagen peptides in the serum has served as a biomarker of fibrosis. Such peptides are procollagen type I carboxy-terminal and procollagen type III amino-terminal peptides, serum type IV collagen, laminin, hyaluronic acid, and YKL-40. These biomarkers have been combined with clinical parameters to form various formulas or indices to predict the severity of fibrosis in disease states such as chronic hepatitis C, nonalcoholic fatty liver disease, and biliary atresia, with variable success.

Liver cancer is one of the dreaded complications of chronic liver disease, and these patients are at risk of developing the malignancy during the course of their illness. Early detection is important as advanced cancer could exclude them from liver transplantation. The classic biomarker of liver malignancy is α-fetoprotein, even though it is not specific to liver tumors, as it may also be raised in chronic hepatitis and cirrhosis. Other serum biomarkers that may be used in conjunction with α-fetoprotein are des-γ-carboxy prothrombin and Golgi protein 73. Elevated cytokines have also been detected, e.g. IL-6, IL-8, and IL-10. Various miRNA profiles have been described that can detect the presence of hepatocellular carcinoma on a background of chronic liver disease.

Anil Dhawan

Baetge EE, Dhawan A, Prentice AM (eds): Next-Generation Nutritional Biomarkers to Guide Better Health Care. Nestlé Nutr Inst Workshop Ser, vol 84, pp 91–102, (DOI: 10.1159/000436992)
Nestec Ltd., Vevey/S. Karger AG., Basel, © 2016

Stratified Medicine: Maximizing Clinical Benefit by Biomarker-Driven Health Care

Sharat Singh

Prometheus Laboratories Inc., San Diego, Calif., USA

Abstract

Stratified medicine involves the use of biomarkers to differentiate patient populations into subsets that provide more detailed information about the specific causes of conditions, and predict how patients will respond to a given drug or combination of drugs. Biomarker-driven patient stratification can empower clinicians by providing accurate assessments of patient status (diagnostic and prognostic utility) in order to strategize treatment planning and delivery (predictive and monitoring utility) based on information extracted from biomarker profiles. This approach may also help presymptomatic individuals by delaying the onset of disease, minimize the severity of the disease, or possibly prevent disease occurrence. Patients and clinicians may benefit from this approach as it may allow for the transformation of current empirical 'medical practice' to efficient data-driven 'individualized therapeutic strategies' with a low risk of medical error. The health care industry may also benefit from this approach by designing clinical trials based on appropriate patient stratification. Here, recent advances in the field of stratified medicine are highlighted in the context of our efforts to integrate this rapidly evolving concept into our research, and to ultimately develop potential diagnostic/prognostic/predictive products and nutritional solutions for individual patients and consumers.

Redefining the Use of Diagnostics in Medicine

Medical practice has always relied upon 'individualizing' treatment strategies based on an individual patient's clinical history and the physician's experience. Traditionally, diagnostic utilities have been used to 'confirm' decisions made by

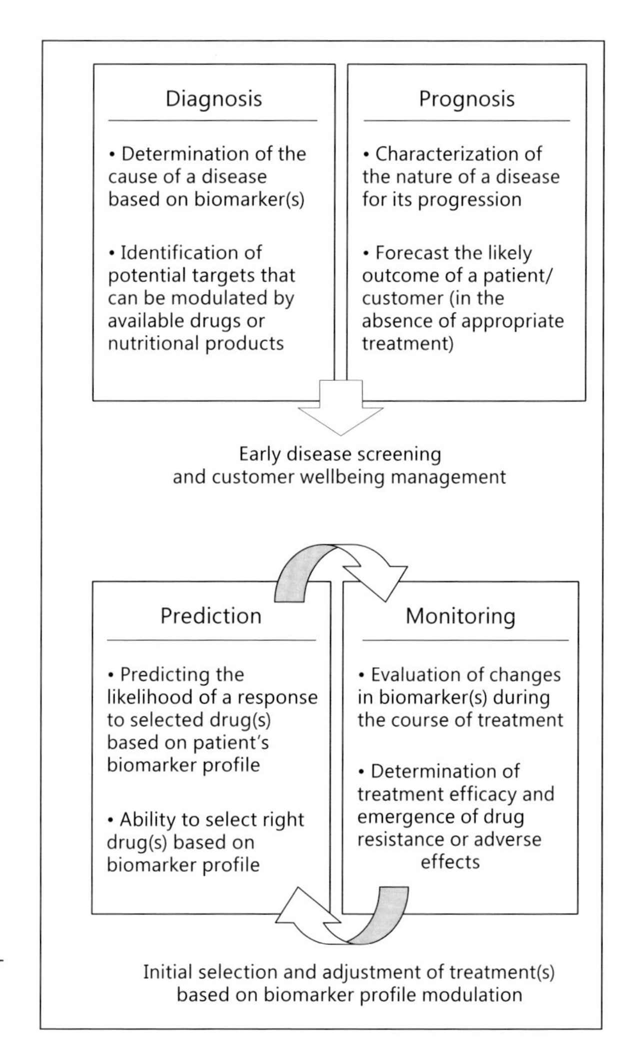

Fig. 1. The use of biomarkers may impact subsequent clinical strategies.

clinicians based on observable symptoms. However, the role of biomarkers in the clinic and in drug development is rapidly expanding due to the advancement of both analytical technologies and our understanding of the molecular basis of diseases. Indeed, recent advances in biomarker-based diagnostics are redefining how diseases should be categorized and/or differentiated.

Diagnostic tools can provide valuable information that is critical to: (1) identifying the cause of a disease, (2) determining the disease course, (3) selecting the most effective treatment option(s) to maximize clinical outcome while minimizing adverse effects, and/or (4) monitoring efficacy during the course of treatment (fig. 1). Accurate assessment of the disease, its natural course of progression along with potential drug targets can influence the physician's engagement strategy.

What Is a Biomarker?

A biomarker is a molecular, anatomical, physiological, histological, or biochemical factor with observable biological characteristics. In order to be effectively utilized in the clinic, biomarkers need to be objectively measurable. Good biomarkers can function as surrogates for the underlying cause(s) of disease as they act as biological indicators of disruption of homeostasis caused by pathological processes. The significance of diagnostics in the health care industry is rapidly increasing due to the accumulation of a repertoire of biomarkers with clear associations to diseases and observable phenotypes.

We are already familiar with the effective use of biomarkers in the context of infectious diseases, where it is critical to identify the pathogenic agent(s) responsible for an infection (e.g. detection of HIV-associated RNA). Often these biomarkers are used to monitor the efficacy of treatment strategies (e.g. viral load in the case of HIV). While it is relatively simple to use biomarkers for infectious diseases, it is often difficult to use them in the clinic for other diseases due to intrinsic complexities associated with pathophysiology. Biomarkers are used to identify the susceptibility of patients for certain diseases such as BRCA-1 for breast cancer. However, due to its incomplete penetrance (68% in breast cancer and 60% in ovarian cancer by age 70), the utility of BRCA-1 in the clinical setting is quite controversial, and the status of this biomarker should be evaluated in the context of each individual [1].

The role of biomarkers in predicting the likelihood of achieving a desired treatment outcome while minimizing adverse effects increasingly resonates with health care providers and other stakeholders, including payers and regulators. Some biomarkers function as surrogates for monitoring the efficacy of given treatments, thereby assisting clinicians in evaluating their clinical strategies. As diseases are reclassified based on their underlying molecular/cellular causes rather than on symptoms or organ-associated phenotypic display, better clinical strategies can be utilized to help prevent or delay disease onset, or minimize disease severity.

Prognostic Use of Biomarkers

Cardiovascular Disease

Statin therapy is effective at reducing cardiovascular (CV) incidents among patients with prior myocardial infarction, stroke, diabetes, or overt hyperlipidemia. It is strongly recommended that lipid-lowering therapies such as statins should be considered in these patients as an adjunct to aggressive lifestyle interventions. However, over 50% of heart attacks and strokes occur among apparently healthy individuals with average or low levels of cholesterol. Hence, there is an unmet need for novel screening of biomarkers to identify individuals at high risk of CV incidents despite the absence of hyperlipidemia in the primary care setting. A prospective study (JUPITER: Justification for the Use of Statin in Prevention: An Intervention Trial Evaluating Rosuvastatin) that enrolled 17,802 healthy individuals with elevated levels of high-sensitivity C-reactive protein (hsCRP) reported a dramatic reduction in CV events with 20 mg/day rosuvastatin compared with placebo [2]. In this study, elevated hsCRP levels were a predictive marker for CV events among apparently healthy persons without hyperlipidemia. Therefore, as a biomarker, hsCRP is *prognostic* for CV incidents and *predictive* of response to rosuvastatin, and its value can be used to screen/monitor apparently healthy individuals at risk of CV events [2].

Cancer
While a single biomarker may provide sufficient information for a certain condition such as the likelihood for CV incidents among individuals with high levels of hsCRP, it often takes a number of biomarkers to be an effective diagnostic tool. Multiple diagnostic applications that rely on panels of biomarkers to determine the severity of a disease as well as the likelihood of response to chemotherapy (e.g. Oncotype DX or MammaPrint for breast cancer) have been utilized in the oncology clinic. Based on current treatment guidelines, estrogen receptor-positive breast cancer patients are treated with hormonal therapy along with chemotherapy. In the TAILORx trial, profiling of 21 gene-based biomarkers was integrated into clinical decision making to facilitate maximal efficacy while minimizing unnecessary adverse effects by limiting chemotherapy to patients who are at higher risk of disease recurrence [3]. The avoidance of unnecessary chemotherapy based on biomarkers also results in cost reductions associated with patient management.

Inflammatory Bowel Disease
There is an incomplete understanding of the pathogenesis of inflammatory bowel disease (IBD), but an algorithm of a combination of serological, genetic, and inflammatory (SGI) biomarkers has been effectively utilized to better differentiate IBD from overlapping symptomatic manifestations. Differentiating IBD into either Crohn's disease or ulcerative colitis impacts treatment strategies caused by underlying differences in immunopathophysiology. Biomarker profiling based on noninvasive testing is preferred, particularly in the pediatric setting, as this allows clinicians to avoid colonoscopy in young patients. One of the most important applications of SGI in patients with established IBD is to determine the severity of disease, and moreover the risk of having complications or rapid disease progression. The conventional approach to managing IBD is to

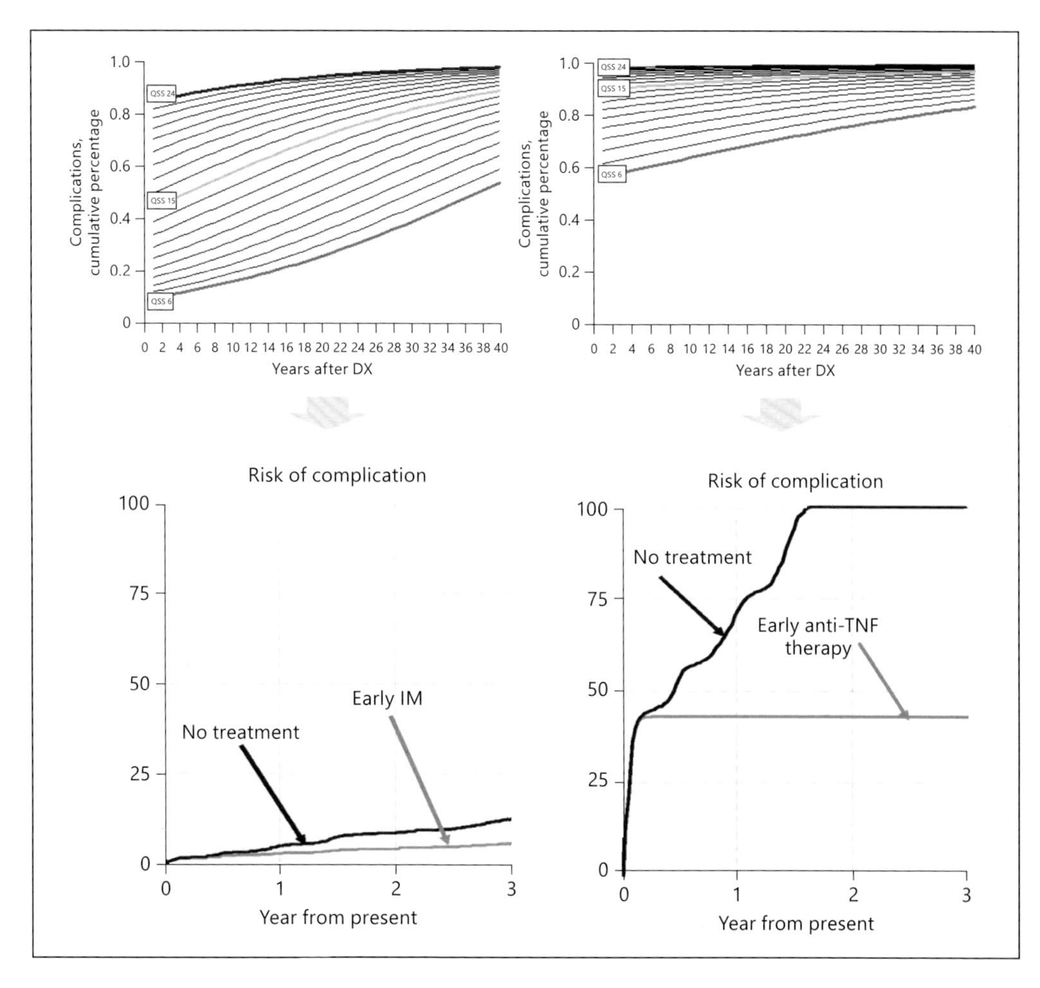

Fig. 2. Principles behind the PROSPECT study: Crohn's disease prognostics (top) and risk of complications (bottom). DX = Diagnosis; IM = immunomodulators.

induce/maintain clinical remission by the progressive intensification of aminosalicylates followed by corticosteroids. However, biologics are used as a last resort in patients with severe disease.

'PROSPECT', a prospective clinical study, utilizes the SGI algorithm to identify high-risk Crohn's disease patients and maximize the clinical response to anti-TNF-α biologics, thereby preventing/slowing disease progression and preventing complications (fig. 2) [4]. Despite an increased number of therapeutic options for IBD, approximately one third of IBD patients require intestinal resection, as the primary focus in the clinic remains the control of symptoms. Hence, there is a critical unmet need for the prognostic identification of individuals at the preclinical stage, when there may be opportunities to alter the natural course of the disease.

Predictive Use of Biomarkers

Prognostic markers provide information regarding the outcome of a disease irrespective of therapy. For example, breast cancer patients with overexpression of human epidermal growth factor receptor 2 (HER2) have a poor clinical outcome. However, with the appropriate therapeutic agent specifically targeting HER2, a status which was once considered only as a prognostic biomarker, its use as a predictive biomarker for HER2-targeting therapies has begun.

Patients with activating epidermal growth factor receptor (EGFR) mutation-positive non-small cell lung cancer (NSCLC) showed dramatic improvement in progression-free survival (PFS) when treated with EGFR-targeting erlotinib versus gemcitabine plus carboplatin in the OPTIMAL study (fig. 3) [5]. This result has strengthened the value of predictive biomarkers, and was further demonstrated in advanced NSCLC patients treated with gefitinib (IPASS: Iressa Pan-Asia Study) [6]. While INTACT (Iressa NSCLC Trial Assessing Combination Treatment) appeared to be unsuccessful when the clinical responses of all enrolled patients were evaluated without stratification, separating patients based on EGFR mutation status demonstrated a significant PFS benefit among patients with EGFR mutation [7, 8]. Furthermore, EGFR-mutant patients showed worse PFS when treated with conventional chemotherapy regimens containing carboplatin plus paclitaxel in IPASS [6]. This outcome confirmed the EGFR mutation as a predictive biomarker for a positive response to EGFR-targeting inhibitors and negative response to conventional chemotherapy. Similarly, in the BRIM-3 (BRAF Inhibitor in Melanoma-3) study, melanoma patients with the BRAF mutation also responded dramatically when treated with the BRAF inhibitor vemurafenib (response rate 48%) versus conventional treatment with dacarbazine (5%; fig. 3). The BRAF mutation is a highly predictive biomarker providing information critical for determining clinical strategies for melanoma patients [9]. This is an example of a number of positive predictive biomarkers in routine clinical use today (table 1).

Most current biomarkers are based on mutations or gene fusion, but only minor subsets of cancers are treated based on these predictive biomarkers. The vast majority of cancers require technologies that are able to decipher the complex pathophysiology responsible for disease occurrence, as well as disease reoccurrence without clear dominant driver mutations. Furthermore, most cancer patients with advanced disease, even those with favorable biomarker profiles, ultimately relapse with recurrent disease, in which the initial targeting agent is no longer effective as the tumor evolves using alternate pathways. Hence, there is an urgent need for methods to identify functional predictive biomarkers to alter treatment strategies in order to maintain control over dynamically evolving disease.

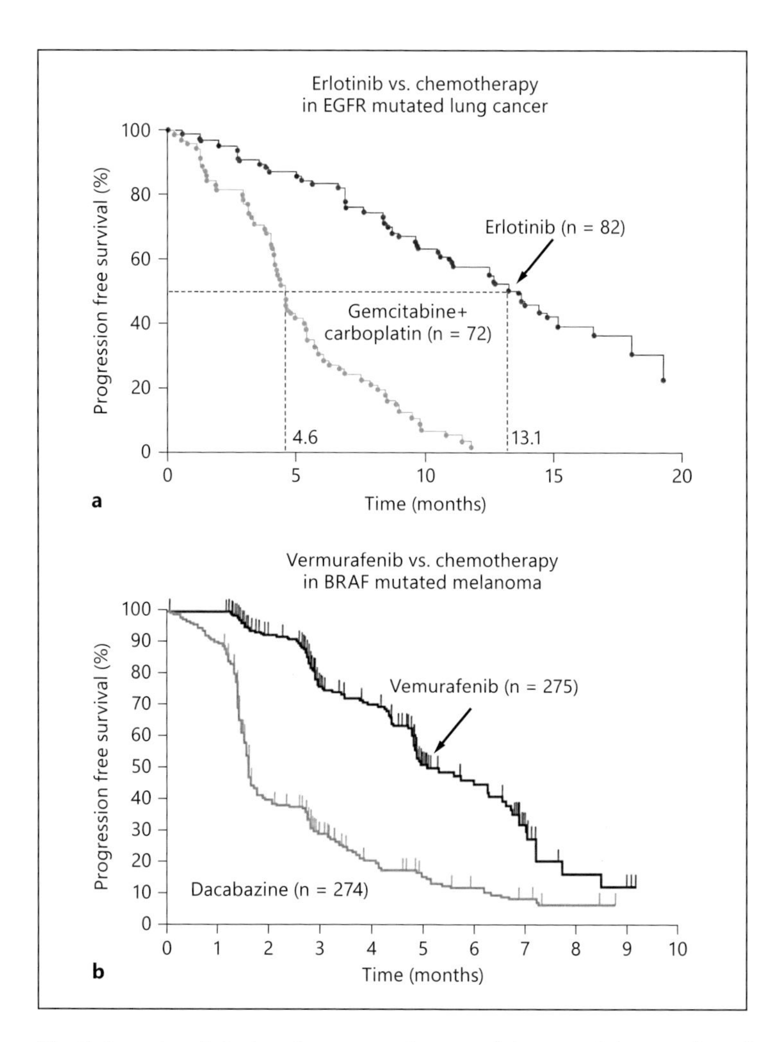

Fig. 3. Superior clinical performance of targeted drugs and the use of predictive biomarkers. **a** OPTIMAL study. **b** BRIM-3 study.

A proximity-mediated immunomicroarray technology (CEER: collaborative enzyme enhanced reactive immunoassay) developed by Prometheus/Nestlé Health Science has the capacity to evaluate small amounts of biological specimens (fig. 4). This technology is able to evaluate the functional status of multiple proteins in several pathways that have the potential to drive tumor proliferation with extreme sensitivity. CEER is based on signal amplification between two detector antibodies with enhanced specificity due to the requirement for three independent epitope-binding events. Rather than evaluating a single biomarker at a time, comprehensive evaluation of pathway physiology using CEER has opened clinical opportunities to keep up with evolving diseases. When NSCLC patients

Table 1. Examples of predictive biomarkers used in oncology clinics

Biomarker	Disease area	Drugs	Diagnostics
HER2	Breast cancer	Trastuzumab, Tykerb	FISH (gene amplification)
HER2	Breast cancer	Trastuzumab, Tykerb	IHC (overexpression)
Oncotype DX	Breast cancer	Chemotherapy	Panel of RNA expression
EGFR	Colorectal cancer	Gefitinib, erlotinib	EGFR mutation panel
BRAF	Melanoma	Vemurafenib, dabrafenib	BRAF mutation
ALK	NSCLC	Crizotinib	AKL-EML gene fusion
KRAS	Colorectal cancer	Erbitux	KRAS mutation
BCR-Abl	CML	Imatinib	BCR-Abl gene fusion
cKIT	GIST	Imatinib	cKIT mutation

CML = Chronic myeloid leukemia; GIST = gastrointestinal stromal tumor.

were evaluated for EGFR as well as other signaling proteins, cMET emerged as an important biomarker in predicting whether an NSCLC patient would respond to EGFR inhibitors with a high level of confidence. In the context of cMET, patients with higher EGFR/cMET index (EGFR-driven tumor) had a better response to EGFR-targeting agents, while patients with lower EGFR/cMET index (cMET-driven tumor) did not benefit from EGFR inhibitor monotherapy. Patients with activated cMET and with activating EGFR mutation would benefit from a combination of EGFR and cMET inhibitors (fig. 4). Patients with high EGFR/cMET index who initially responded well to EGFR inhibitors often developed resistance to the treatment. Upon analyzing tumor cells obtained from pleural effusion, we identified insurgence pathway signaling often triggered by activation of cMET and other receptor tyrosine kinases [10]. CEER could be a model platform to identify 'functional predictive biomarkers' for keeping up with evolving disease as tumors develop resistance. Similarly, 434 gastric cancer patients were profiled for the activation status of multiple receptor tyrosine kinases (RTKs, HER1, HER2, p95HER2, HER3, cMET, IGF1R, and PI3K). The results revealed that the combined RTK activity index was a good prognostic composite biomarker for predicting disease-free survival [11]. Furthermore, treatment monitoring by evaluating circulating tumor cells, as well as tumor cells from ascitic fluids, demonstrated a clinical potential for the rapid adaption of combinatorial treatments for patients with advanced gastric cancer.

Predictive biomarkers are often accurate for single disease indications. While BRAF mutation status works well as a predictive biomarker for BRAF inhibitor treatment in melanoma patients, the same mutation found in colorectal cancer is not predictive for clinical response when treated with BRAF inhibitor. Based on a multiplexed pathway analysis approach, several competing/redundant pathway drivers were identified in colorectal cancer patients. Multiplexed bio-

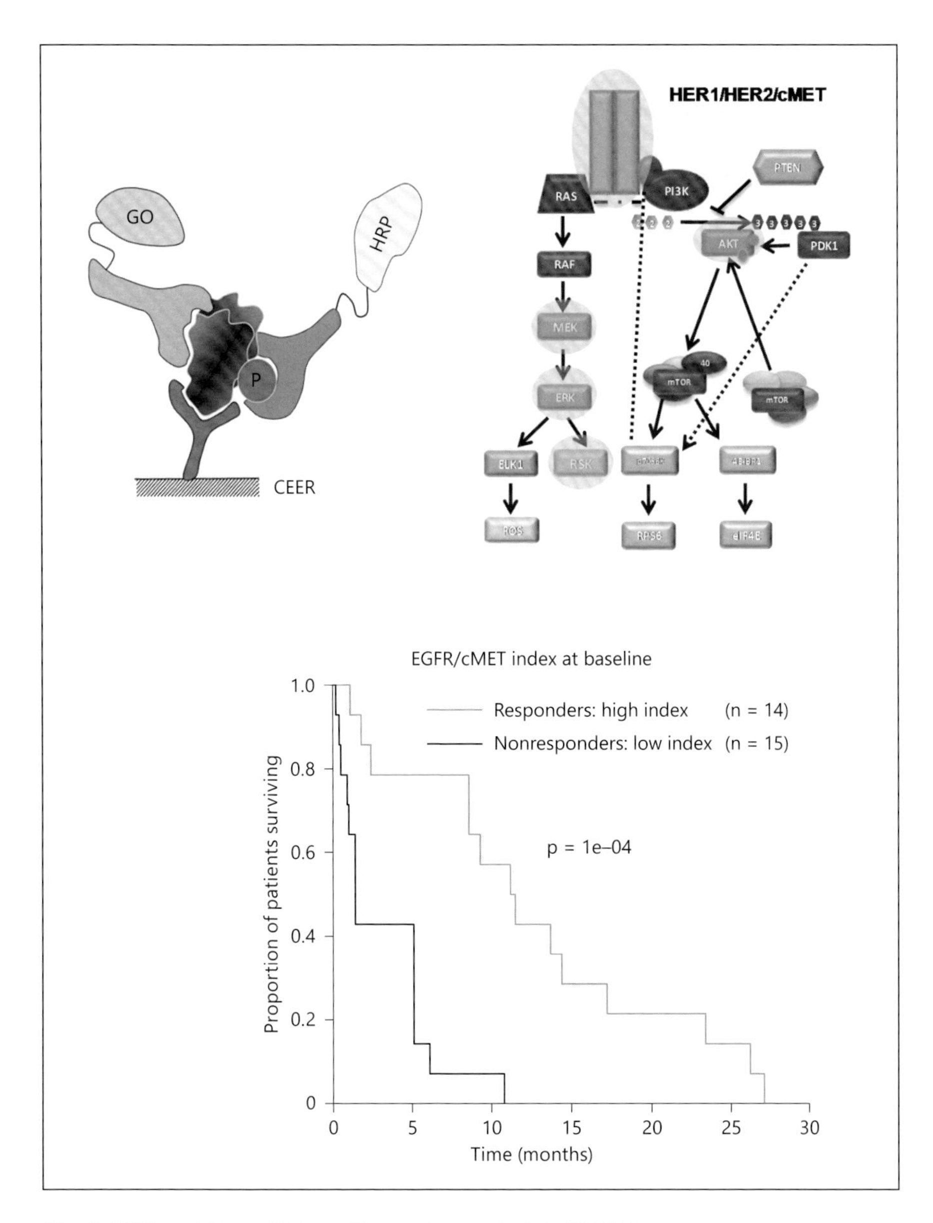

Fig. 4. CEER and its multiplexed biomarker analysis in NSCLC.

marker analysis delivered more accurate and comprehensive pathway status information to better select patients who would respond to a specific targeting agent and further provided the additional strategy for combining multiple agents. Tumors that have metastasized to the brain are often driven by different proliferative mechanisms compared with primary brain tumors; therefore, systemic treatments are often not effective in controlling metastatic brain tumors.

The sensitivity and specificity of CEER allows treatment options to be selected based on the predictive biomarker profile determined from small numbers of tumor cells isolated from cerebral spinal fluid, with dramatic clinical responses [12]. This study, along with others that evaluated tumor cells isolated from the circulation, ascitic fluids, and pleural effusions, demonstrated the value of clinical strategies based on predictive biomarkers.

Dealing with evolving disease is not limited to the field of oncology, but is also critical in chronic diseases such as IBD. Patients with advanced IBD are treated with monoclonal antibodies targeting TNF-α such as infliximab, which is often highly effective in the management of refractory IBD. However, more than one third of patients do not respond to induction therapy (primary nonresponse), and even among initial responders, the response wanes over time in 20–60% of patients (secondary nonresponse) [13]. Current hypotheses for the causes of nonresponse include: (1) inadequate serum levels or rapid consumption of the drug due to high inflammatory disease burden and (2) development of immunogenicity against biologics or activation of a non-TNF-α pathway [13].

The homogeneous mobility shift assay was developed by Prometheus/Nestlé Health Science to overcome the clinical challenges of determining the level of therapeutic biologic in the presence of antibodies to a drug. Recent studies using this test have demonstrated clinical advantages over other assays and have shown how monitoring can improve patient outcomes in various settings (e.g. loss of response or drug holiday) [14]. The development of antibodies against biologics (antibodies to infliximab) increases the probability of active disease even at low concentrations, and hence evaluating the level of antibodies to infliximab is critical to identifying alternative clinical strategies [15]. Patients with no antibody against biologic and with sufficient amount of biologic in the serum who still have active disease need further evaluation. Of note, some patients have active disease despite seemingly adequate anti-TNF-α levels in the serum; this is possibly due to high TNF-α levels in locally inflamed tissue resulting in a neutralizing effect on the anti-TNF-α drug [16].

It is not practical to subject patients with active disease to endoscopic evaluation; therefore, the need to identify biomarkers that function as surrogates for determining the severity of damaged mucosa, monitor the effectiveness of treatments, and provide an achievable therapeutic target for 'mucosal healing' is urgent. The pilot use of multipurpose biomarker panels was recently reported where biomarkers associated with the wound healing process were indicative of sustained mucosal healing [17]. The expanded second-generation panel is currently being validated, and will provide evidence-driven induction and proactive monitoring.

Conclusions

The value of diagnostics compared to therapeutics has been historically underappreciated by the health care system. However, the stratification of patients for specific diseases has gained tremendous momentum in the postgenomic era. Using therapeutics in patient groups with minimal benefit, and often with severe adverse events, taxes the entire health care system. In order to provide highly effective health care solutions, patients should be stratified for specific diseases:

- Disease stratification is essential to better understand disease diversity/subtypes.
- Patient stratification is critical to strategize the best treatment options for each individual.
- Stratified health care will reduce the number of adverse drug reactions.
- Treatment of complex diseases requires comprehensive biomarker panels that can keep up with evolving disease.
- In order to become a leader in 'emerging' markets, we must utilize technologies and informatics to translate 'research/clinical findings and learnings' into practical products that address unmet needs in the clinic.
- Stratified medicine will reduce ineffective prescriptions and patient management leading to cost reductions for the health care system.
- Prognostic biomarkers may identify individuals at risk even in the presymptomatic stage, providing opportunities for nutritional therapeutics to prevent further disease progression.
- Stratification is desirable for every entity in the health care system: patients, care provides, payers, and pharmaceutical and diagnostic companies.
- While stratification may narrow the pool of eligible patients for a given indication, it can improve market share and should enhance efficacy with minimal side effects.
- Biomarker-based treatment monitoring will maximize clinical benefit and will lead to better patient compliance.

Disclosure Statement

The author declares that no financial or other conflict of interest exists in relation to the contents of the chapter.

References

1 Evans DG, Shenton A, Woodward E, et al: Penetrance estimates for BRCA1 and BRCA2 based on genetic testing in a clinical cancer genetics service setting: risks of breast/ovarian cancer quoted should reflect the cancer burden in the family. BMC Cancer 2008;8: 155.

2 Ridker PM, Danielson E, Fonseca FA, et al: Rosuvastatin to prevent vascular events in men and women with elevated C-reactive protein. N Engl J Med 2008;359:2195–2207.

3 Sparano JA: The TAILORx trial: individualized options for treatment. Commun Oncol 2006;3:494–496.

4 Siegel CA, Siegel LS, Hyams JS, et al: Real-time tool to display the predicted disease course and treatment response for children with Crohn's disease. Inflamm Bowel Dis 2011;17:30–38.

5 Zhou C, Wu YL, Chen G, et al: Erlotinib versus chemotherapy as first-line treatment for patients with advanced EGFR mutation-positive non-small-cell lung cancer (OPTIMAL, CTONG-0802): a multicentre, open-label, randomised, phase 3 study. Lancet Oncol 2011;12:735–742.

6 Fukuoka M, Wu YL, Thongprasert S, et al: Biomarker analyses and final overall survival results from a phase III, randomized, open-label, first-line study of gefitinib versus carboplatin/paclitaxel in clinically selected patients with advanced non-small-cell lung cancer in Asia (IPASS). J Clin Oncol 2011;29: 2866–2874.

7 Giaccone G, Herbst RS, Manegold C, et al: Gefitinib in combination with gemcitabine and cisplatin in advanced non-small-cell lung cancer: a phase III trial – INTACT 1. J Clin Oncol 2004;22:777–784.

8 Bell DW, Lynch TJ, Haserlat SM, et al: Epidermal growth factor receptor mutations and gene amplification in non-small-cell lung cancer: molecular analysis of the IDEAL/INTACT gefitinib trials. J Clin Oncol 2005; 23:8081–8092.

9 Chapman PB, Hauschild A, Robert C, et al: Improved survival with vemurafenib in melanoma with BRAF V600E mutation. N Engl J Med 2011;364:2507–2516.

10 Ahn M-J, Park S, Sun J-M, et al: Roles of cMET/ErbB3 activation and overexpression in the development of resistance to EGFR inhibitors in NSCLC patients. Abstract ASCO Annu Meet, 2013.

11 Lee J, Kim S, Kim P, et al: A novel proteomics-based clinical diagnostics technology identifies heterogeneity in activated signaling pathways in gastric cancers. PLoS One 2013; 8:e54644.

12 Pastorino S, Pingle SC, Langley E, et al: Cerebrospinal fluid pharmacokinetics and pharmacodynamics following high-dose erlotinib treatment in brain cancer patients. AACR Annu Meet, San Diego, 2014.

13 Ordás I, Feagan BG, Sandborn WJ: Therapeutic drug monitoring of tumor necrosis factor antagonists in inflammatory bowel disease. Clin Gastroenterol Hepatol 2012;10: 1079–1087.

14 Wang SL, Hauenstein S, Ohrmund L, et al: Monitoring of adalimumab and antibodies-to-adalimumab levels in patient serum by the homogeneous mobility shift assay. J Pharm Biomed Anal 2013;78–79:39–44.

15 Vande Casteele N, Khanna R, Levesque BG, et al: The relationship between infliximab concentrations, antibodies to infliximab and disease activity in Crohn's disease. Gut 2014, DOI: 10.1136/gutjnl-2014-307883.

16 Yarur AJ, Jain A, Sussman DA, et al: The association of tissue anti-TNF drug levels with serological and endoscopic disease activity in inflammatory bowel disease: the ATLAS study. Gut 2015, DOI: 10.1136/gutjnl-2014-308099.

17 de Bruyn M, Bessissow T, Billiet T, et al: Biomarker panel for prediction of mucosal healing in patients with Crohn's disease under infliximab therapy. 9th Congr ECCO, Copenhagen, 2014.

Baetge EE, Dhawan A, Prentice AM (eds): Next-Generation Nutritional Biomarkers to Guide Better Health Care. Nestlé Nutr Inst Workshop Ser, vol 84, pp 103–110, (DOI: 10.1159/000436993)
Nestec Ltd., Vevey/S. Karger AG., Basel, © 2016

The Gut Microbiome, Its Metabolome, and Their Relationship to Health and Disease

Gary D. Wu

Division of Gastroenterology, Perelman School of Medicine, University of Pennsylvania, Philadelphia, PA, USA

Abstract

Despite its importance in maintaining the health of the host, growing evidence suggests that gut microbiota may also be an important factor in the pathogenesis of various diseases. The composition of the microbiota can be influenced by many factors, including age, genetics, host environment, and diet. There are epidemiologic data associating diet with the development of inflammatory bowel disease as well as evidence that diet can influence both the form and the function of the microbiome. Based on this evidence, studies are now underway to examine the effect of defined formula diets, an effective therapeutic modality in Crohn's disease, on both the gut microbiome and its metabolome as a therapeutic probe. Diet has an impact upon both the composition and the function of the microbiota in part through small-molecule production that may influence the development of both immune-mediated and metabolic diseases. By comparing dietary intake, the gut microbiota, and the plasma metabolome in omnivores versus vegans, we provide evidence that the production of certain bacterial metabolites is constrained by the composition of the gut microbiota. In total, these results demonstrate the potential promise of dietary manipulation of the gut microbiota and its metabolome as a modality to both maintain health and treat disease.

Introduction

The human gut microbiome is one of the most densely populated bacterial communities on Earth with up to 10^{11} organisms per gram of fecal weight composed of over a 1,000 species, most of which are obligate anaerobes [1]. The bacterial concentration, as well as complexity, increases proximally from the stomach and duodenum from approximately 10^2 to 10^3 aerobic organisms per gram luminal content to 10^{11} to 10^{12} organisms per gram luminal content distally, where anaerobic organisms predominate in the cecum and colon. Throughout, the collective genome of the bacteria is 150-fold greater than that of its human host [2]. Indeed, humans should be viewed as biologic 'supraorganisms' that are dynamic and carry out functions in parallel or cooperatively. Most gut microbes are obligate anaerobes, many of which are fastidious and difficult to grow in vitro, making traditional culture techniques of limited value in characterizing the composition of the gut microbiota. The development of culture-independent methods, mainly through the use of high-throughput DNA sequencing, has provided new means to evaluate the gut microbiome and its relationship to inflammatory bowel disease (IBD). There are two primary methods that utilize deep-sequencing technologies to characterize the microbiome. The first approach uses small-subunit ribosomal RNA (16S rRNA) gene sequences (for Archaea and bacteria), or 18S rRNA gene sequences (for eukaryotes) as stable phylogenetic markers to define the lineages present in a sample [3]. Another approach uses shotgun metagenomic sequencing that permits the characterization of both the structure and the genomic representation of the microbial community. This broad-based genomic community evaluation helps to elucidate the functions encoded by the genomes of the gut microbiota. Additional technologies, such as metatranscriptomics and metaproteomics, may also provide a deeper understanding of microbial function through the direct evaluation of gene expression.

Mammalian hosts have coevolved to exist with their gut microbiota in a mutualistic relationship whereby we provide a uniquely suited environment in return for physiological benefits provided to us by our gut microbiota [4]. Examples of the latter include the fermentation of indigestible carbohydrates to produce short-chain fatty acids that are utilized by the host, biotransformation of conjugated bile acids, synthesis of certain vitamins, degradation of dietary oxalates, hydrolysis of urea by urease activity that participates in host nitrogen balance, and education of the mucosal immune system [4]. Nevertheless, there is growing evidence that the gut microbiota is associated with a number of diseases, particularly in animal model systems, but, potentially, also in humans. In this review, I will explore the relationship between diet, the bacterial gut microbiome, its metabolome, and their possible relationships to health and disease.

Inflammatory Bowel Disease as a Paradigm for the Dysbiotic Microbiota

The alterations in the gut microbiome that are associated with IBD are often described as being 'dysbiotic', or having an altered community structure, implying that there is a functional imbalance between enteric bacteria with potentially pathogenic influences and bacteria who have a benign or beneficial effect on the host [5]. Other than the effectiveness of fecal microbiota transplantation for the treatment of recurrent *Clostridium difficile* infection [6], there is currently no clear evidence to confirm this notion in humans. An alternative explanation is that the observed alteration in the gut microbiome of patients with IBD is simply a consequence of the intestinal inflammatory response without consequence to the host. Growing evidence suggests that the altered composition of the dysbiotic microbiota is an adaptive response of a complex microbial community to environmental stress imposed by the intestinal inflammatory process leading to the production of electron acceptors or an increase in oxidative stress perpetuating the growth of more oxygen-tolerant organisms that belong to the Proteobacteria and Actinobacteria phyla [7, 8]. There is, however, evidence for a functional effect of a 'dysbiotic' intestinal microbiota in animal models [9]. Together, these studies suggest a causal role for the dysbiotic microbiota in perpetuating the chronicity of intestinal inflammation in patients with IBD.

Modulation of the Dysbiotic Microbiota for Health – The Role of Diet

Since the gut microbiota unquestionably plays a critical role in the pathogenesis of IBD, certain nongenetic factors associated with the development of IBD may be due, in part, to their effects on the gut microbiota. Environmental factors that may alter the composition of the gut microbiota include diet, the consumption of antibiotics and other xenobiotics, and geographic location. Population-based studies suggest that IBD is unevenly distributed throughout the world, with the highest disease rates occurring in industrialized nations. One theory, the hygiene hypothesis, suggests that humans living in more industrialized countries are exposed to an altered microbial environment with less complex microbial communities at an early age leading to the development of an immune system less able to 'tolerate' exposure to the microbial-laden environment in later life resulting in inappropriate immune activation. Consistent with this notion is the possible role of diet in light of the differences in access to clean water and availability of food refrigeration in underdeveloped parts of the world. Alternatively, a 'westernized' diet rich in animal fat and protein, while low in fiber, may alter the gut microbiome in a way that increases the risk for the development of IBD. Indeed, reduced richness of the gut microbiota has been shown to be associated with multiple inflammation-associated diseases and the consumption of a westernized diet relative to that found in residents of more agrarian cultures, where diets are primarily plant based [10, 11].

There are reasonable data to support a role for diet in IBD pathogenesis. Several investigators have examined the association of dietary patterns and the incidence of IBD [12, 13]. For example, the authors of a systematic review concluded that high dietary intake of total fats, polyunsaturated fatty acids, ω-6 fatty acids, and meat were associated with an increased risk of Crohn's disease and ulcerative colitis; high fiber and fruit intakes were associated with a decreased risk of Crohn's disease, and high vegetable intake was associated with a decreased risk of ulcerative colitis [13]. These studies support a potential role for dietary patterns in the pathogenesis of IBD.

In Crohn's disease, exclusive enteral nutrition with elemental, semi-elemental, and defined formula diets have been widely studied for the induction of remission and are considered the first-line therapy in certain parts of the world [14, 15]. These diets are also efficacious in maintaining remission [16]. Despite the efficacy of this therapeutic modality, the mechanisms by which exclusive enteral nutrition reduces inflammation in patients with Crohn's disease are unknown. Current studies are underway to determine the effect of exclusive enteral nutrition on the composition of the gut microbiota in the hope of identifying microbial taxa and/or metabolites that are either beneficial or deleterious in Crohn's disease pathogenesis. Conceptually, of fundamental importance to these studies is to understand how the consumption of these defined formulas are different from dietary intake of whole foods – are they providing something beneficial not delivered in whole foods or are they preventing the consumption of something deleterious in whole foods that is not present in a defined formula diet?

Diet as a Substrate for Metabolite Production and Its Influence on Health and Disease

Although diet can have an effect on the composition and/or richness of the gut microbiota, perhaps more important is its impact on the microbial metabolome. Indeed, diet may serve as a substrate that can be used by the gut microbiota for the production of small molecules that, after first-pass metabolism through the liver, can have an important impact on host physiology [17]. One example would be the role that gut microbiota may play in augmenting the development of atherosclerosis through the production of certain metabolites of dietary lipid phosphatidylcholine that are associated with the risk for the development of cardiovascular disease. Using a targeted approach to identify plasma metabolites which predict the cardiovascular risk in patients, Wang et al. [18] and Tang et al. [19] identified a novel pathway linking dietary lipid intake, intestinal microbiota, and atherosclerosis. Foods rich in phosphatidylcholine are a major source of choline. Catabolism of choline by the intestinal microbiota results in the formation of the gas TMA (trimethylamine) that is metabolized by the liver to form TMA oxide

(TMAO), a small molecule that is strongly associated with the increased risk for coronary vascular disease in humans. TMAO also augments the development of atherosclerosis in animal models, thus providing the first link between dietary lipid intake, intestinal microbiota, and the risk for the development of atherosclerosis [18]. A similar pathway has been identified for the conversion of dietary carnitine, which is abundant in red meat, and its conversion into TMAO [20]. Recently, the bacterial gene family responsible for the conversion of choline into TMA, known as choline TMA-lyases, has been described [21] where investigators have shown that the greatest abundance of nonpathogenic bacterial taxa with this gene representation are located in three of the four major phyla of the human gut microbiome. With this knowledge, several possible avenues can be envisioned by which this information can now be used to develop technologies that may directly impact upon human health.

Another example, which is perhaps more relevant to the pathogenesis and, possibly, the treatment of immune-mediated diseases such as IBD, would be the delivery of indigestible carbohydrates to the gut microbiota through dietary intake leading to the production of short-chain fatty acids by bacterial fermentation that play a role in immune function [22, 23], intestinal hormone production, and lipogenesis [24]. Exploiting this relationship between diet and the gut microbiota is a strategy to treat immune-mediated diseases by restoring immune tolerance by the activation of regulatory T cells. For example, additional prebiotics could deliver fermentable substrates to an enhanced microbiota fortified by 'next-generation' probiotics designed to alter the composition of the gut microbiota to produce greater levels of short-chain fatty acids that would, in turn, activate specific G-protein-coupled receptors to decrease inflammation. An alternative approach would be to develop a small molecule to directly target the activation of specific G-protein-coupled receptors known to modulate immune function [22].

Impact of Diet on the Composition of the Human Microbiota and Its Production of Metabolites

Prior studies in animal models and globally distinct human populations focusing on a 'westernized' versus an agrarian plant-based diet suggest that the impact of diet on taxonomy may be large, implying that the gut microbiota plays a significant role in health [25–27]. By contrast, more moderate dietary interventions, which can be sustained in humans on the long term, suggest that the impact of diet may be more modest [28]. To examine the impact of a plant-based diet within the context of a westernized environment, we compared measures of dietary intake, gut microbiota composition, and the plasma, urinary, and fecal metabolome of healthy human subjects consuming either a long-term vegan or omnivorous diet [29]. Vegans consumed significantly less micronutrients associated with protein and fat intake than omnivores, whereas carbohydrate consumption was significantly greater. The plasma metabolome of vegans differed markedly from that of omnivores, but the gut microbiota was surprisingly similar. Unlike prior studies of individuals living in agrarian societies [26], the higher consumption of fermentable substrate in vegans was not associated with higher levels of fecal short-chain fatty acids, a finding confirmed in a 10-day controlled feeding experiment [28]. Similarly, the proportion of vegans capable of producing equol, a soy-based gut microbiota metabolite, was less than reported in Asian societies despite the high consumption of soy-based products. The notion that differences in the composition of the gut microbiota in globally distinct cultures might impact upon function is also supported by the identification of a specific gut bacterial species capable of seaweed algal metabolism found in the gut microbiota of the Japanese but not in residents of North America [30]. These results suggest that environmental factors that shape the composition of the gut microbiota may have a significant effect on the relationship between substrates provided to the gut microbiota through dietary consumption of the host and the production of microbial metabolites.

Conclusion

There is growing interest in targeting the gut microbiota as a strategy to maintain health and treat disease. Advances in DNA sequencing and mass spectroscopy technologies together with enhanced biocomputational tools designed to analyze high-dimensional data sets have identified microbial taxa and metabolites that could be developed as therapeutic strategies to target specific disease states through the development of next-generation pre-, pro-, and synbiotics. The effect of diet on both the composition of the gut microbiota and its production of metabolites is likely to be a very important component of this strategy. Additional studies are needed to characterize environmental factors independent of diet which may play a critical role in shaping the composition of the gut microbiota in globally distinct human societies which, in turn, may have an effect on the production of beneficial metabolites, such as short-chain fatty acids, from the diet and gut microbiota. The development of prebiotics to deliver substrates for the gut microbiota to produce desirable metabolites that will favor health may need to take differences in the composition of the gut microbiota in various societies around the world into consideration.

Disclosure Statement

The author has received research funding and honoraria from Nestle for organizing and/or speaking in scientific symposia.

References

1 Uhlig HH, Powrie F: Dendritic cells and the intestinal bacterial flora: a role for localized mucosal immune responses. J Clin Invest 2003;112:648–651.
2 Xu J, Gordon JI: Inaugural article: honor thy symbionts. Proc Natl Acad Sci U S A 2003; 100:10452–10459.
3 Marchesi JR: Prokaryotic and eukaryotic diversity of the human gut. Adv Appl Microbiol 2010;72:43–62.
4 Hooper LV, Gordon JI: Commensal host-bacterial relationships in the gut. Science 2001;292:1115–1118.
5 Tamboli CP, Neut C, Desreumaux P, et al: Dysbiosis in inflammatory bowel disease. Gut 2004;53:1–4.
6 van Nood E, Vrieze A, Nieuwdorp M, et al: Duodenal infusion of donor feces for recurrent *Clostridium difficile*. N Engl J Med 2013; 368:407–415.
7 Winter SE, Lopez CA, Baumler AJ: The dynamics of gut-associated microbial communities during inflammation. EMBO Rep 2013; 14:319–327.
8 Albenberg L, Esipova TV, Judge CP, et al: Correlation between intraluminal oxygen gradient and radial partitioning of intestinal microbiota. Gastroenterology 2014;147: 1055–1063.e8.
9 Garrett WS, Lord GM, Punit S, et al: Communicable ulcerative colitis induced by T-bet deficiency in the innate immune system. Cell 2007;131:33–45.
10 Cotillard A, Kennedy SP, Kong LC, et al: Dietary intervention impact on gut microbial gene richness. Nature 2013;500:585–588.
11 Le Chatelier E, Nielsen T, Qin J, et al: Richness of human gut microbiome correlates with metabolic markers. Nature 2013;500: 541–546.
12 Chapman-Kiddell CA, Davies PS, Gillen L, et al: Role of diet in the development of inflammatory bowel disease. Inflamm Bowel Dis 2010;16:137–151.
13 Hou JK, Abraham B, El-Serag H: Dietary intake and risk of developing inflammatory bowel disease: a systematic review of the literature. Am J Gastroenterol 2011;106:563–573.
14 Sandhu BK, Fell JM, Beattie RM, et al: Guidelines for the management of inflammatory bowel disease in children in the United Kingdom. J Pediatr Gastroenterol Nutr 2010; 50(suppl 1):S1–S13.
15 Caprilli R, Gassull MA, Escher JC, et al: European evidence based consensus on the diagnosis and management of Crohn's disease: special situations. Gut 2006; 55(suppl 1):i36–i58.
16 Takagi S, Utsunomiya K, Kuriyama S, et al: Effectiveness of an 'half elemental diet' as maintenance therapy for Crohn's disease: a randomized-controlled trial. Aliment Pharmacol Ther 2006;24:1333–1340.
17 Holmes E, Li JV, Marchesi JR, et al: Gut microbiota composition and activity in relation to host metabolic phenotype and disease risk. Cell Metab 2012;16:559–564.
18 Wang Z, Klipfell E, Bennett BJ, et al: Gut flora metabolism of phosphatidylcholine promotes cardiovascular disease. Nature 2011; 472:57–63.
19 Tang WH, Wang Z, Levison BS, et al: Intestinal microbial metabolism of phosphatidylcholine and cardiovascular risk. N Engl J Med 2013;368:1575–1584.
20 Koeth RA, Wang Z, Levison BS, et al: Intestinal microbiota metabolism of L-carnitine, a nutrient in red meat, promotes atherosclerosis. Nat Med 2013;19:576–585.
21 Craciun S, Balskus EP: Microbial conversion of choline to trimethylamine requires a glycyl radical enzyme. Proc Natl Acad Sci U S A 2012;109:21307–21312.
22 Maslowski KM, Vieira AT, Ng A, et al: Regulation of inflammatory responses by gut microbiota and chemoattractant receptor GPR43. Nature 2009;461:1282–1286.

23 Smith PM, Howitt MR, Panikov N, et al: The microbial metabolites, short-chain fatty acids, regulate colonic Treg cell homeostasis. Science 2013;341:569–573.
24 Samuel BS, Shaito A, Motoike T, et al: Effects of the gut microbiota on host adiposity are modulated by the short-chain fatty-acid binding G protein-coupled receptor, Gpr41. Proc Natl Acad Sci U S A 2008;105:16767–16772.
25 Ley RE, Hamady M, Lozupone C, et al: Evolution of mammals and their gut microbes. Science 2008;320:1647–1651.
26 De Filippo C, Cavalieri D, Di Paola M, et al: Impact of diet in shaping gut microbiota revealed by a comparative study in children from Europe and rural Africa. Proc Natl Acad Sci U S A 2010;107:14691–14696.
27 Muegge BD, Kuczynski J, Knights D, et al: Diet drives convergence in gut microbiome functions across mammalian phylogeny and within humans. Science 2011;332:970–974.
28 Wu GD, Chen J, Hoffmann C, et al: Linking long-term dietary patterns with gut microbial enterotypes. Science 2011;334:105–108.
29 Wu GD, Compher C, Chen EZ, et al: Comparative metabolomics in vegans and omnivores reveal constraints on diet-dependent gut microbiota metabolite production. Gut 2014, DOI: 10.1136/gutjnl-2014-308209.
30 Hehemann JH, Correc G, Barbeyron T, et al: Transfer of carbohydrate-active enzymes from marine bacteria to Japanese gut microbiota. Nature 2010;464:908–912.

Baetge EE, Dhawan A, Prentice AM (eds): Next-Generation Nutritional Biomarkers to Guide Better Health Care. Nestlé Nutr Inst Workshop Ser, vol 84, pp 111–119, (DOI: 10.1159/000436995)
Nestec Ltd., Vevey/S. Karger AG., Basel, © 2016

The Scientific Challenge of Expanding the Frontiers of Nutrition

Serge Rezzi · Soren Solari · Nicolas Bouche · Emmanuel E. Baetge

Nestlé Institute of Health Sciences, Lausanne, Switzerland

Abstract

Nutritional research is entering a paradigm shift which necessitates the modeling of complex interactions between diet, genetics, lifestyle, and environmental factors. This requires the development of analytical and processing capabilities for multiple data and information sources to be able to improve targeted and personalized nutritional approaches for the maintenance of health. Ideally, such knowledge will be employed to underpin the development of concepts that combine consumer and medical nutrition with diagnostic targeting for early intervention designed to maintain proper metabolic homeostasis and delay the onset of chronic diseases. Nutritional status is fundamental to any description of health, and when combined with other data on lifestyle, environment, and genetics, it can be used to drive stratified or even personalized nutritional strategies for health maintenance and preventive medicine. In this work, we will discuss the importance of developing new nutrient assessment methods and diagnostic capabilities for nutritional status to generate scientific hypotheses and actionable concepts from which to develop targeted and eventually personalized nutritional solutions for health protection. We describe efforts to develop algorithms for dietary nutrient intake and a holistic nutritional profiling platform as the basis of understanding the complex nutrition and health interactome.

Introduction

The human body is a complex and dynamic network of physiological regulatory and adaptive processes that is constantly under modulation from multiple intrinsic (genetic, metagenetic, and metabolic) and extrinsic (lifestyle and environment) factors. Scientific evidence continues to grow on the pivotal role of nutrition in maintaining health as well as in delaying the onset of chronic disorders. Indeed, the extension of human life expectancy and the pandemia of chronic and multifactorial disorders such as obesity, type-2 diabetes, and various neurological disorders are naturally positioning nutrition at the forefront of sustainable approaches for healthy aging and preventive medicine. This is logically understood as nutritional intake is the process that exposes the human body to a multitude of organic and inorganic substances contained in food and beverages on a chronic basis. Food and beverages provide the water, macronutrients, and micronutrients necessary to sustain the growth, metabolism, and repair of biological systems. Several so-called essential nutrients, such as vitamins, minerals, fatty acids, and amino acids, required for normal human body function are dependent on exogenous sources from dietary origins, for example, due to the inherent inability to either biosynthesize those nutrients at all or at a scale that does meet the body's metabolic requirements. Moreover, some nutrients can also be conditionally essential when their relative biosynthetic capacity is lower than the metabolic demand due to specific life stages or disease conditions.

Several eating habits, such as diet rich in saturated fats and trans fats, are known to be associated with an increased risk of cardiometabolic disorders. Moreover, there is growing scientific evidence regarding the links between nutrients and complex multifactorial diseases [1]. It is thus probable that people's expectations from dietary and nutrient recommendation guidelines will continue to grow together with the increasing scientific knowledge between nutrition and health. Reference recommendations have been established to provide dietary values evaluated as sufficient to meet nutrient requirements in a population group [2]. However, due to the lack of a standardized approach for determining nutrient recommendations at the global scale, the terminology, sometimes referring to dietary reference values, recommended dietary allowances, recommended nutrient intakes, nutrient intake values, or dietary reference intakes, and the reference values for particular nutrients vary substantially between countries [3]. Reference nutrient values are usually defined at the population scale and are largely based on scientific associations between dietary intake and health or clinical outcomes. For instance, in Europe and North America, dietary reference intake is calculated from the average requirement of a single nutrient incremented by twice its standard deviation in order to cover 97.5% of the general population. This calculation assumes a statistically normal distribution for the nutrient of interest. Thus, whilst these reference systems have enabled guidance for consumer and regulatory authorities for optimal nutrient intake, one should acknowledge that they are based on reductionist scientific methodology, i.e. one nutrient being considered at a time.

From Reductionist to 21st-Century System Nutrition

Nutrition is a multifactorial process as nutrients are neither absorbed nor metabolized one at a time but as a complex mixture of thousands of chemical entities delivered from foods and beverages. Nutrient-nutrient or other interactions can occur during the digestion, which subsequently modulate nutrient absorption and bioavailability [4]. The iron and ascorbic acid (vitamin C) interaction is a typical example showing that the latter enhances the intestinal absorption of non-heme iron from the diet [5]. On the other hand, genetic polymorphisms are increasingly recognized as important determinants of the biology of nutrients in humans. The known mutation in the gene encoding for 5,10-methylenetetrahydrofolate reductase, considered a genetic factor for developing vascular disease, illustrates this gene-nutrient axis with a reduced enzymatic activity that associates with impaired levels of homocysteine [6]. Another example has recently been reported for the genetic variation in the vitamin D binding protein and circulating levels of vitamin D metabolites following vitamin D supplementation. Results showed that genetic variation in the vitamin D binding protein was associated with a different nutrition status, which was measured using circulating levels of 25(OH)D vitamers, following supplementation with vitamin D_3 but not with vitamin D_2 [7]. Levels of several liposoluble vitamins such as α-tocopherol, γ-tocopherol, α-carotene, β-carotene, lycopene, β-cryptoxanthin, lutein, and zeaxanthin could also be related to variants of genes involved in lipid metabolism [8]. Genome-wide association studies revealed that variants in transferrin and hemochromatosis genes significantly associated with serum transferrin levels [9]. When considering genetic variability in mammalian superorganisms, one should also consider the contribution of the extended genome carried by symbiotic partners such as the gut microbiota. Indeed, the gut microbiota interacts with the host via a broad spectrum of metabolic reactions, including processes involved in nutrient digestion and energy recovery from foods. The gut microbiota is also reported to be involved in the production of vitamins, such as vitamin K (menaquinone) and group B vitamins (biotin, pyridoxine, cobalamin, riboflavin, folates, thiamine, and nicotinic and pantothenic acids) [10]. Consequently, understanding interindividual peculiarities in nutrient status should also consider the possible variability in the gut microbiota composition and function.

Thus, isolating the biological effect of a single nutrient does not reflect what happens in a highly complex network of nutrient metabolic interactions that can occur in a mammalian superorganism such as the human being. This complexity may explain in part the current lack of molecular knowledge about nutrient and micronutrient acquisition, as well as their metabolism, compartmentalization across different cell types, transport, and mechanism of action on specific cellular phenotypes. Therefore, one of the most important scientific challenges of the 21st century is to understand the way dietary nutrients are taken up, stored, mobilized, and utilized by the body both at the system and at the cellular levels. Such a scientific challenge can be addressed through a paradigm shift which combines the conventional reductionist approach with an integrated systems approach where the nutrient-body interactome can be properly captured with its inherent complexity. Similar to postgenomic sciences such as metabonomics and proteomics, systems nutrition is expected to provide new insights into understanding the interindividual nutritional peculiarities with the aid of high-resolution phenotypic profiling that would be based on comprehensive measurements of the nutrient status together with other clinical parameters. Nutritional phenotypes could be leveraged to better define nutrient requirements for groups of people sharing nutrient metabolic phenotypes, in shared environments with similar genetic backgrounds. The outcomes of this research are thus expected to pave the way towards personalized nutrition to maintain health and prevent disease [1].

The Path Towards the Next Generation of Targeted and Personalized Nutrition

A principal scientific challenge to enable systems nutrition is to make the multiple components of what is known as the nutritional status more standardized. High-resolution nutritional profiling can then be used as the hallmark of modern nutrition: to identify unmet nutrient needs, and guide dietary and/or lifestyle adaptations to meet specific nutrient demands, and finally as a monitoring platform to measure the success of personalized programs. Nutritional status combines multiple measurements ranging from the evaluation of nutrient intake, anthropometric and clinical assessments to the quantitative analysis of nutrient levels, or some of their related status and functional biomarkers, in biological samples such as blood and urine.

The choice of diet results from a decision process that integrates educational, cultural, and socioeconomic factors, taste and palatability perception, and known biological determinants such as food intolerance or allergy. Moving towards personalized nutrition will require a greater empowerment of people's dietary choices and their culinary and cultural habits so that specific nutrient requirements are fulfilled. Moving towards personalized nutrition for health and wellness requires a paradigm shift from today's nutrition towards empowerment of people on nutrition for health and wellness (fig. 1). This implies the development of novel holistic solutions for people that will assist them in decisions

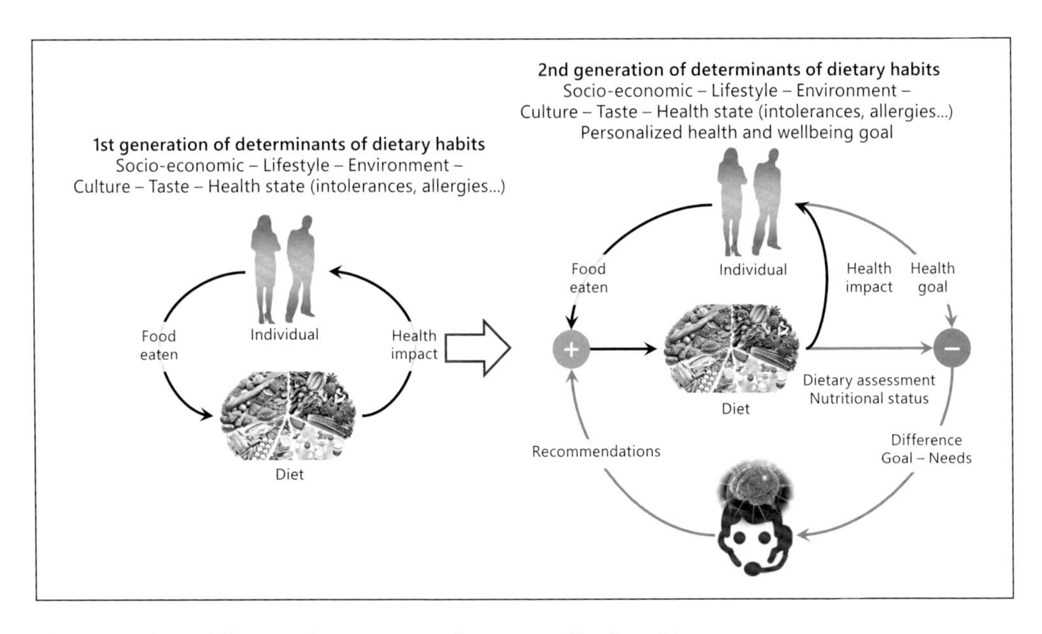

Fig. 1. Paradigm shift towards next-generation personalized nutrition.

for dietary adjustments to match their specific nutrient needs. To attain their ultimate goal for maintenance of optimal nutritional health, such solutions must be holistic, convenient, sustainable, and proven to be biologically effective for improvement of the nutritional status. The holistic aspect will require the discovery and integration of convenient and accurate nutrient intake and nutritional status assessments together with nutritional service and counseling offers as well as novel ways to produce, combine, and process whole foods and/or enhance their nutrient density with personalized nutrient fortification. A feedback loop to the person on his/her nutrient intake and nutritional status should be provided to demonstrate nutritional benefits of personalization through an intelligent recommendation engine (fig. 1). Empowerment of people would therefore require the integration of the three following components:

(1) Capability to intelligently assess the habits and nutrient density of the people's diet

(2) Capability to effectively measure the nutritional status of people to identify unmet nutrient needs

(3) Innovative mechanisms to deliver the needed or differential nutritional recommendations

A convenient and holistic tool to assess people's dietary habits and nutrient intake is currently being developed as the Nutrition Health Concept (NHC). The NHC provides values relative to conventional dietary questionnaires through a method of scoring consumables, e.g. food/meals/diets, using a data-driven ap-

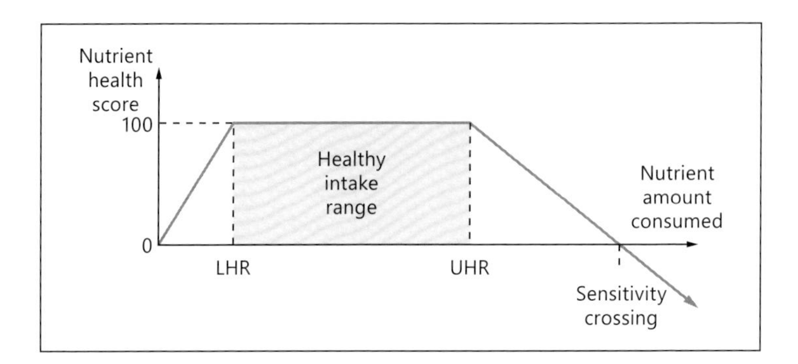

Fig. 2. An NHC score is designed by defining the nutrients to be included in the score, the lower (LHR) and the upper healthy range (UHR) and sensitivity crossing for each nutrient (as above) and the weights to perform a weighted average across all nutrient scores. A perfect NHC score = 100 is achieved when all individual nutrient health scores are within the consumed healthy range. Any nutrient or energy outside their healthy range will decrease the score depending on how far away from the healthy range the value is.

proach. The NHC defines nutritional health as consuming nutrients (broadly defined as the measurable components in food) within lower and upper healthy ranges. Those ranges are defined for each nutrient, including macro- and micronutrients, over a set period of time (i.e. 1 day or 1 week). Unlike other dietary assessment tools that classify nutrients into 'good/bad', 'qualifying/disqualifying', or 'need to have more of/need to limit' categories, the NHC does not depend on such a preclassification of foods. Basically, if all nutrients and even energy intake are given healthy ranges, then all nutrients and energy can be treated equally. A healthy range rather than a category determines the definition of nutrient health for the individual. As displayed in figure 2, each nutrient in an NHC score has a healthy range of consumption with both a lower healthy range and upper healthy range. Therefore, the NHC enables people to generate their nutrient and health score based upon their personal dietary choices. A perfect NHC score is indicated as 100 and is achieved when all individual nutrient health scores are within the consumed healthy range. One can then combine a set of foods so that the amount of all nutrients consumed fall within the targeted healthy range. As shown in figure 2, consumption outside the healthy range (low or high) will result in lower scores.

The design of the NHC provides thus a personalized scoring system that can be used to guide dietary recommendations to either maintain or enhance the health score to its optimal value. There is no singularly defined NHC score, but rather an infinite number of potential NHC scores that can be created based on the individual's goals. As an example, a NHC score may be designed for a diabetic person who is trying to lose weight. In this case, the weight and sensitivity to the nutrient, sugar, might both be set high, and the caloric intake range lower, which will drive nutritional recommendations accordingly. In another case, an individual may require more iron and protein consumption over a specified time and an NHC score can be designed to reflect the individual's needs and goals. When combined with the measurement of the individual's nutritional status, the NHC becomes an even more powerful dietary recommendation tool, because it takes the individual biological uniqueness of measured blood-based nutrient levels together with the diet consumed into account. With the feedback loop of nutritional status readouts, the NHC algorithm can then be further fine-tuned so that dietary choices can be optimized according to the individual nutritional response to specific dietary choices and nutrient intakes.

The field of biological analysis of the nutritional status is marked by the high number of different methodological protocols, which are often dedicated and optimized to either single or limited numbers of nutrients. Yet, technological advances in the field of analytics in both separation sciences, i.e. chromatography/ detectors and mass spectrometry, open up unprecedented possibilities to resolve many nutrients in a single run while covering a fairly broad dynamic range of concentrations. Attempts to develop universal and standardized methodologies to quantitatively profile a spectrum of nutrients remain sporadic. Profiling of 16 trace elements in human serum was recently reported using inductively coupled plasma mass spectrometry [11]. Midttun et al. [12] recently published a method to quantify 10 molecular species related to vitamins B_6 and B_2 in human plasma using liquid chromatography coupled to tandem mass spectrometry. Priego Capote et al. [13] developed a method based on a similar technology to measure a series of 10 molecular species related to liposoluble vitamins and their metabolites in human serum. Nevertheless, the nutrition field has not benefited from a so-called 'one-stop shop' comprehensive profiling of nutrients encompassing minerals and trace elements, liposoluble and hydrosoluble vitamins, amino acids, fatty acids, and their related metabolites. This can be achieved using the centralization of multinutrient profiling methods. Such a nutrient profiling platform is foreseen to be a key enabler to modern systems nutrition by enhancing research possibilities to capture interindividual differences in nutrient status as a function of the genetic background, dietary habits, and health status (fig. 3). Applied in clinical studies, the analytical outcomes of such a platform will more holistically supply the nutrient determinants that may explain the degree of individual response to dietary or nutritional intervention, for example, thus opening up the way to identify more systematically specific nutritional requirements associated with health, disease, and environmental conditions.

In the future, it is expected that technological solutions for the biological analysis of nutritional status will evolve towards miniaturization enabling translation either into point-of-care diagnostics or, more long term, into wearable sensor

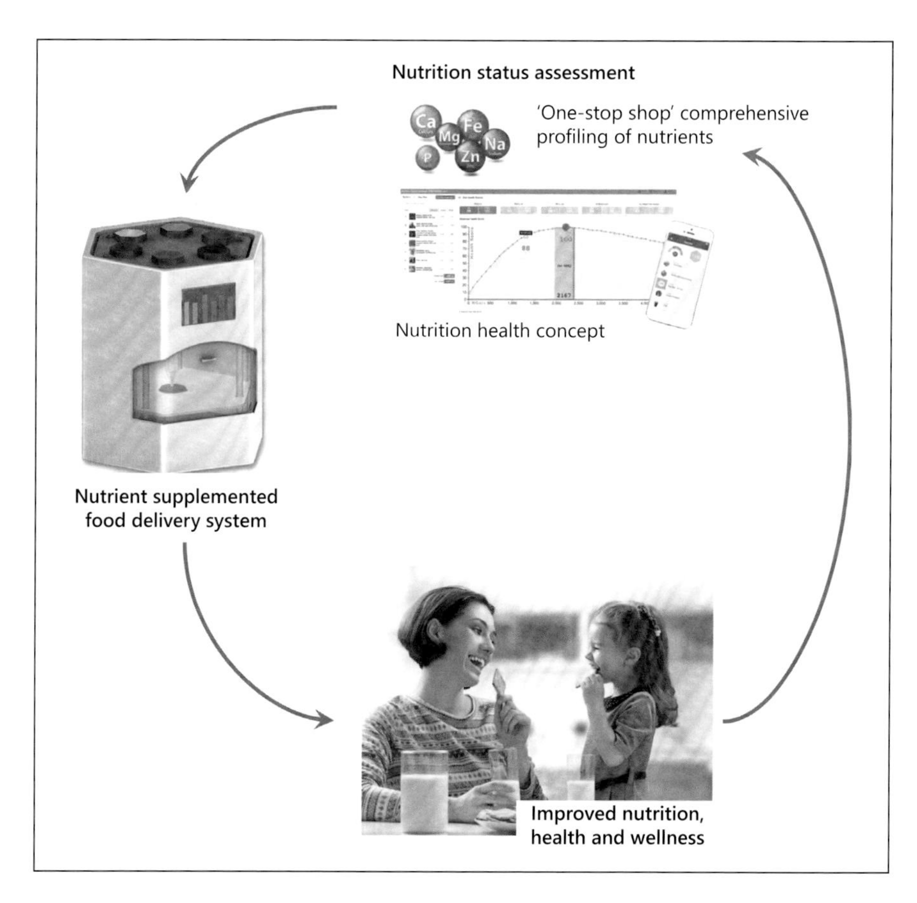

Fig. 3. The next generation of a comprehensive nutrient profiling platform for a personalized nutritional status.

technologies that will enable real-time feedback to people with information about their nutritional status. Technological evolution is also expected in the field of new food and nutrient delivery systems that will offer convenience and nutritional personalization of food and beverages to promote greater personalization of nutritional health with the longer-term ambition of slowing disease onset. The empowerment of nutritional health through application of simplified profiling systems has great promise for human health maintenance.

Disclosure Statement

S. Rezzi, E. Baetge and N. Bouche are employees of Nestlé Institute of Health Sciences and members of the Nestlé group; S. Solari is a former employee of Nestlé Institute of Health Sciences and member of the Nestlé group.

References

1 Rezzi S, Collino S, Goulet L, Martin FP: Metabonomic approaches to nutrient metabolism and future molecular nutrition. TrAC 2013;52:112–119.

2 Pavlovic M, Prentice A, Thorsdottir I, et al: Challenges in harmonizing energy and nutrient recommendations in Europe. Ann Nutr Metab 2007;51:108–114.

3 Van't Veer P, Grammatikaki E, Matthys C, et al: EURRECA-framework for aligning micronutrient recommendations. Crit Rev Food Sci Nutr 2013;53:988–998.

4 Sandström B: Micronutrient interactions: effects on absorption and bioavailability. Br J Nutr 2001;85:S181–S185.

5 Hallberg L, Brune M, Rossander L: Effect of ascorbic acid on iron absorption from different types of meals. Studies with ascorbic-acid-rich foods and synthetic ascorbic acid given in different amounts with different meals. Hum Nutr Appl Nutr 1986;40:97–113.

6 Frosst P, Blom HJ, Milos R, Goyette P, et al: A candidate genetic risk factor for vascular disease: a common mutation in methylenetetrahydrofolate reductase. Nat Genet 1995; 10:111–113.

7 Nimitphong H, Saetung S, Chanprasertyotin S, et al: Changes in circulating 25-hydroxyvitamin D according to vitamin D binding protein genotypes after vitamin D_3 or D_2 supplementation. Nutr J 2013;12:39.

8 Borel P, Moussa M, Reboul E, et al: Human plasma levels of vitamin E and carotenoids are associated with genetic polymorphisms in genes involved in lipid metabolism. J Nutr 2007;137:2653–2659.

9 Benyamin B, McRae AF, Zhu G, et al: Variants in TF and HFE explain approximately 40% of genetic variation in serum-transferrin levels. Am J Hum Genet 2009;84:60–65.

10 Hill MJ: Intestinal flora and endogenous vitamin synthesis. Eur J Cancer Prev 1997;6:S43–S45.

11 Forrer R, Gautschi K, Lutz H: Simultaneous measurement of the trace elements Al, As, B, Be, Cd, Co, Cu, Fe, Li, Mn, Mo, Ni, Rb, Se, Sr, and Zn in human serum and their reference ranges by ICP-MS. Biol Trace Elem Res 2001; 80:77–93.

12 Midttun O, Hustad S, Solheim E, et al: Multianalyte quantification of vitamin B6 and B2 species in the nanomolar range in human plasma by liquid chromatography-tandem mass spectrometry. Clin Chem 2005;51: 1206–1216.

13 Priego Capote F, Jiménez JR, Granados JM, de Castro MD: Identification and determination of fat-soluble vitamins and metabolites in human serum by liquid chromatography/triple quadrupole mass spectrometry with multiple reaction monitoring. Rapid Commun Mass Spectrom 2007;21:1745–1754.

Baetge EE, Dhawan A, Prentice AM (eds): Next-Generation Nutritional Biomarkers to Guide Better Health Care. Nestlé Nutr Inst Workshop Ser, vol 84, pp 121–122, (DOI: 10.1159/000436996)
Nestec Ltd., Vevey/S. Karger AG., Basel, © 2016

Summary on Future Horizons

Section three profiles three separate approaches to stratifying patients and people with respect to disease or health status using antibody, microbial, and nutritional profiling approaches. In the first presentation, *Sharat Singh* highlighted how to use predictive biomarker profiles for the diagnosis, prognosis, prediction, and monitoring of the disease status or biological drug levels in patients and provided examples thereof. He stressed the use of two novel technologies, CEER (collaborative enzyme enhanced reactive immunoassay) and HMSA (homogeneous mobility shift assay) for the dynamic profiling of tumor phenotype and anti-TNF drug levels and autoantibodies, respectively. These novel diagnostic platforms have been applied in both oncology and gastrointestinal disease profiling and monitoring, and have resulted in successful product developments and clinical applications for anti-TNF therapies in inflammatory bowel disease. The platforms have wide application possibilities for the detection of any protein therapeutic or the ultrasensitive detection of peptide antigens.

In the second presentation, *Gary D. Wu* underscored the enormous complexity of the gut microbiome (150 times more genetic information and 10 times more cells) and the clear association of dysbiosis (altered gut microbial communities) with inflammatory bowel disease. Careful phenotyping of the microbial communities and their metabolic signatures can inform us about the state of intestinal health. He reviewed the role of the diet and its influence on the microbial composition, and what metabolic signatures can be measured that correlate or are specifically attributed to altered microbial communities. Furthermore, the role of the diet in association with the living environment can imply distinct functional attributes of the microbiome. Finally, he emphasized that the combination of genomic, metabolomics, dietary, and informatics approaches are necessary to properly diagnose microbial dysbiosis associated with disease and for maintaining microbial homeostasis regarding health.

Finally, in the third presentation, *Emmanuel E. Baetge* discussed the development of a further set of methodologies describing approaches related to nutritional status profiling and dietary diagnosis. These relate to complete nutritional descriptions through the implementation of a one-stop shop for measuring serum profiles of all essential nutrients. These include the essential water- and lipid-soluble vitamins, essential amino acids, fatty acids, and metal ions required for daily health maintenance. It is clear that for improved nutritional knowledge and development of important nutritional medicine, we must begin to systematically apply such technologies. Currently, a clinically validated platform for the measurement of all essential nutrients and their important metabolites is not found anywhere in the world. The development of novel targeted nutritional approaches, or medical nutrition profiling, is obligatory to be able to derive the data needed to better understand and translate the effects of the nutritional status on health maintenance or development of disease. Finally, a simpler means for tracking and personalizing the total nutritional status of the diet over any time frame, together with the capability to make recommendations, is required to begin to empower people and patients to maintain nutritional health. NHC, a Nutrition Health Concept application, was developed to utilize existing data bases of food and nutrition information such that for a given set of calories consumed, a nutritional score from 1 (bad) to 100 (good) can be obtained. This moves us away from good versus bad foods, and more toward personalized complete diet/nutrition that can be adapted to needs or requirements over time and in ranges specified as appropriate. Ultimately, the recommendations for completing one's healthy nutritional status can be coupled to nutritional delivery systems for diet supplementation and to serum-based nutritional profiling for validation and personalized adjustment.

Emmanuel E. Baetge

Subject Index

Acute liver failure, *see* Liver disease
Adiponectin, atherosclerosis marker 84, 85
AIDS, *Salmonella enterica* association in Africa 39, 40
Atherosclerosis
  pathology 81, 82
  vulnerability markers
    adaptive immunity 83, 84
    adiponectin 84, 85
    endothelial microparticles 86
    inflammation 82, 83
    innate immunity 84
    intraplaque hemorrhage 85
    mast cells 84
    myeloperoxidase 84, 85
    pentraxin-3 84, 85
    prospects for study 86, 87
    S100A8/9 84

Bioinformatics
  drug discovery and repurposing 42–44
  mechanistic insights and biomarkers 39–41
  overview 35–38
  prospects 44, 45
  systems biology tools 38, 39
Biomarkers, *see also specific biomarkers*
  definition 93
  descriptive terms 7
  diagnostics 91, 92
  predictive use 96–101
  prognostic use 93–96
  systems-level design 8, 9
Biomarkers of Nutrition for Development (BOND) 17, 18
BMI, *see* Body mass index
Body mass index (BMI), preterm birth phenotype 73
BOND, *see* Biomarkers of Nutrition for Development
BRAF, cancer mutation 96, 98, 99

CEER, *see* Collaborative enzyme-enhanced reactive immunoassay
CF, *see* Cystic fibrosis
Chemokines
  families 89
  liver fibrosis markers 52
  liver inflammation markers 51, 52
Collaborative enzyme-enhanced reactive immunoassay (CEER) 97–99, 121
C-reactive protein (CRP)
  cardiovascular disease marker 94
  nonalcoholic fatty liver disease marker 54
CRP, *see* C-reactive protein
Cystic fibrosis (CF)
  CFTR status and hyperinflammation 42–44
  DNA microarray studies 37
Cytokeratins
  CK-18 as liver cell death marker 50, 51
  hepatocellular carcinoma markers 53

Diabetes type 2, systems
interventions 29, 30

EMPs, *see* Endothelial microparticles
Endothelial microparticles (EMPs),
atherosclerosis markers 86
Epigenetics
global health biomarkers 19–22
overview 3
preterm birth phenotype 77
Erythroferrone, iron status marker 66

Ferritin, nonalcoholic fatty liver disease
marker 54
Food, Drug, and Cosmetic Act 2

Gc-globulin, acute liver failure
marker 53, 54
Global health
epigenetic biomarkers 19–22
micronutrient deficiencies 15, 16
Gut microbiome
dietary influences
inflammatory bowel disease 105, 106
metabolite production
106–108
microbiota in vegans versus
omnivores 107, 108
inflammatory bowel disease
studies 104–106
overview 103, 104

HCC, *see* Hepatocellular carcinoma
Hepatocellular carcinoma (HCC),
markers 52, 53
Hepcidin
iron regulation 60, 61
iron status biomarker 17, 63–65, 67
HER2, cancer prognosis marker 96
HMGB1
drug-induced injury marker 55
liver cell death marker 51
Hyperlipidemia, preterm birth
phenotype 75
Hypertension, preterm birth
phenotype 76

IBD, *see* Inflammatory bowel disease
Inflammation and Nutritional Science for
Programs/Policies and Interpretation
of Research Evidence (INSPIRE) 17
Inflammatory bowel disease (IBD)
gut microbiome studies 104–106
prognostic biomarkers 94–96
refractory disease 100
INSPIRE, *see* Inflammation and Nutritional
Science for Programs/Policies and
Interpretation of Research Evidence
Insulin sensitivity, preterm birth
phenotype 75
INTACT study 96
Iron status
biomarkers
erythroferrone 66
hepcidin 17, 63–65, 67
prospects 67, 68
transferrin receptor 64
clinical applications 62
hepcidin regulation 60, 61
iron supplementation guidance 62, 63
iron-refractory iron deficiency
anemia 64
screen-and-treat program
implementation 65, 66

LECT2, acute liver failure marker 54
Liver disease
biomarkers
acute liver failure 53, 54
cancer 52, 53
cell death 50, 51
disease progression 50
drug-induced injury 55
fibrosis 52
inflammation 51, 52
microRNA 55
nonalcoholic fatty liver disease 54, 55
overview of pediatric disease 49, 50

Mast cells, atherosclerosis markers 84
Metabolomics
gut microbiome, *see* Gut microbiome
preterm birth phenotype 76, 77

Microbiome, *see* Gut microbiome
MicroRNA, liver disease markers 55, 56
MPO, *see* Myeloperoxidase
Myeloperoxidase (MPO), atherosclerosis marker 84, 85

NHC, *see* Nutrition Health Concept
Non-small cell lung cancer (NSCLC), prognostic markers 96–99
Nonalcoholic fatty liver disease, *see* Liver disease
NSCLC, *see* Non-small cell lung cancer
Nutrition Health Concept (NHC) 115–117, 122

PAR, *see* Population-attributable risk
Pentraxin-3, atherosclerosis marker 84, 85
Personalized nutrition, prospects 114–118
Population-attributable risk (PAR) 2
Preterm birth
  adult phenotype 72
  biomarkers
    body composition 73–75
    body mass index 73
    cardiovascular system 76
    epigenetics 77
    hyperlipidemia 75
    hypertension 76
    insulin sensitivity 75
    metabolomics 76, 77
  determinants of phenotype 72–74
  epidemiology 71, 72
  prospects for study 78, 79
  sex differences in outcomes 77, 78
PROSPECT study 95

Reductionism, supplantation by systems-level design 4–6, 113, 114

S100A8/9, atherosclerosis markers 84
Salmonella enterica, AIDS association in Africa 39, 40
Systems flexibility
  interventions for optimization 29–31
  market implications 31, 32
  optimal health association 26, 27
  stress response biomarkers 27–29
  timeline of health trajectory 31
Systems-level design
  biomarkers 8, 9
  human heterogeneity 6, 8
  reductionism supplantation 4–6, 113, 114

TFBS analysis, *see* Transcription factor binding site analysis
TLRs, *see* Toll-like receptors
TMA, *see* Trimethylamine
Toll-like receptors (TLRs), atherosclerosis role 84
TRAIL, liver cell death marker 51
Transcription factor binding site (TFBS) analysis 36, 41
Transferrin receptor, iron status marker 64
Trimethylamine (TMA), gut microbe synthesis 106, 107

Vitamin A, BOND initiative biomarkers 17, 18